Better Spending for Localizing Global Sustainable Development Goals

This book centres around an intense debate among donors, policymakers, development practitioners, and academics on the efficacy of aid in eradicating poverty while promoting human development.

It seeks to fill the gap in present literature by presenting stories of better spending through implementing Sustainable Development Goals and addressing Agenda 2030 via indigenization of global development goals with initiatives at local and national levels. The book adopts an innovative approach to dealing with aid effectiveness by highlighting the relevance of better spending, rather than excessive spending. It does so with real-life examples of interventions made in the Global South to realize the vision of "thinking globally and acting locally." These case studies speak to the significance of communities' role in shouldering responsibility for planning, financing, operating, and maintaining local developmental initiatives. The examples also demonstrate how aid serves its purpose when used as an investment in communities and enterprising individuals, in order to realize the strategic impact of giving and build a local "receiving mechanism" for indigenizing and achieving global development goals.

The book references cases of better spending by governments, philanthropists, and civil society organizations (CSOs) from across Asia, Africa, and Latin America on a range of issues and will, thus, be of interest to development practitioners, policymakers, donors, philanthropists, civil society organizations, and academics and students of international development studies.

Fayyaz Baqir is a visiting scholar at the University of Ottawa. He served as Senior Advisor on civil society at the United Nations, and CEO of Trust for Voluntary Organizations. He received top contributors' awards from UNDP's Global Poverty Reduction Network.

Nipa Banerjee has a PhD from the University of Toronto in development studies and has served for over 40 years as an international development practitioner, policy analyst and advisor, and professor, at different periods of her career. Her experience includes 34 years with CIDA, Canada's former ODA agency, notably as head of mission in Thailand, Cambodia, Laos, and Afghanistan and as Counselor Development representing CIDA in Bangladesh, Indonesia and India.

Sanni Yaya is Full Professor of economics and global health, and Director and Associate Dean of the School of International Development and Global Studies. His work focuses on a broad array of multidisciplinary topics in development and global health.

The Dynamics of Economic Space
Series Editor: Nuri Yavan
Ankara University, Turkey

The IGU Commission on The Dynamics of Economic Space aims to play a leading international role in the development, promulgation and dissemination of new ideas in economic geography. It has as its goal the development of a strong analytical perspective on the processes, problems, and policies associated with the dynamics of local and regional economies as they are incorporated into the globalizing world economy. In recognition of the increasing complexity of the world economy, the Commission's interests include industrial production; business, professional, and financial services, and the broader service economy including e-business; corporations, corporate power, enterprise, and entrepreneurship; and the changing world of work and intensifying economic interconnectedness.

Making Connections
Technological Learning and Regional Economic Change
Edited by Edward J. Malecki and Päivi Oinas

Regional Change in Industrializing Asia
Regional and Local Responses to Changing Competitiveness
Edited by Leo van Grunsven

Interdependent and Uneven Development
Global – Local Perspectives
Edited by Michael Taylor and Sergio Conti

The Industrial Enterprise and its Environment
Spatial Perspectives
Edited by Sergio Conti, Edward J. Malecki and Päivi Oinas

Towards Coastal Resilience and Sustainability
Edited by C. Patrick Heidkamp and John Morrissey

Better Spending for Localizing Global Sustainable Development Goals
Examples From the Field
Edited by Fayyaz Baqir, Nipa Banerjee and Sanni Yaya

For more information about this series, please visit: www.routledge.com/ The-Dynamics-of-Economic-Space/book-series/ASHSER1030

Better Spending for Localizing Global Sustainable Development Goals

Examples From the Field

Edited by
Fayyaz Baqir, Nipa Banerjee,
and Sanni Yaya

Routledge
Taylor & Francis Group

LONDON AND NEW YORK

First published 2020
by Routledge
2 Park Square, Milton Park, Abingdon, Oxon OX14 4RN

and by Routledge
52 Vanderbilt Avenue, New York, NY 10017

Routledge is an imprint of the Taylor & Francis Group, an informa business

First issued in paperback 2021

British Library Cataloguing-in-Publication Data
A catalogue record for this book is available from the British Library

Library of Congress Cataloging-in-Publication Data
A catalog record for this book has been requested

ISBN: 978-0-367-33845-9 (hbk)
ISBN: 978-1-03-208761-0 (pbk)
ISBN: 978-0-429-32232-7 (ebk)

Typeset in Times New Roman
by Apex CoVantage, LLC.

Dedicated to young minds striving to beat scarcity in the age of abundance

Contents

List of figures	x
List of tables	xii
Foreword	xiii
List of contributors	xv

PART I
Citizens, state, and markets   1

1   Introduction and background: global goals, local action   3
   FAYYAZ BAQIR

PART II
Frameworks for better spending   17

2   Anatomy of an effective development operation: a finance
   minister considers whether to borrow from the World Bank   19
   ANTHONY CHOLST

3   SDG tracking through SSC for better SDG spending   44
   HASANUZZAMAN ZAMAN AND SYED SAJJADUR RAHMAN

PART III
**South Asian landscape: moving from patronage
to participation**   61

4   Society for Elimination of Rural Poverty (SERP):
   an Indian experience of rapid poverty reduction through
   women's empowerment   63
   SHOAIB SULTAN KHAN

5 An inspirational story: galvanizing local action for realizing
global development goals: the story of BRAC making the
21st-century drive to eradicate extreme poverty at local levels   80
NIPA BANERJEE

6 Water management is water measurement   95
FAYYAZ BAQIR

7 Redefining and localizing development in Pakistan   107
FURQAN ASIF

8 Market-led development   121
SHAKEEL AHMAD

PART IV
Resilient communities in fragile states: Central South Asia   135

9 Community-driven development as a mechanism for realizing
global development goals: the National Solidarity Programme
and Citizens' Charter Afghanistan Program   137
NIPA BANERJEE

10 Multidimensional poverty measurement and aid efficiency:
a case study from Afghanistan   151
ABDULLAH AL MAMUN AND SANNI YAYA

PART V
Non-zero options for local development   165

11 Development aid and access to water and sanitation in
sub-Saharan Africa   167
SANNI YAYA, OGOCHUKWU UDENIGWE, AND HELENA YEBOAH

12 From more spending to better spending: the case of food and
nutrition security in Ethiopia   186
SANNI YAYA, NEVILLE SUH, AND RICHARD A NYIAWUNG

13 Don't spend more, spend better: improving the social efficiency
of water and sanitation services in Uruguay and South Africa   201
ADRIAN MURRAY AND SUSAN SPRONK

PART VI
Global South in the Global North   215

14 **Indigenous peoples in Canada – the case of Global South
in the Global North**   217

PART VII
Social capital: the highest form of capital   219

15 **Conclusion: can a single spark create a prairie fire?**   221
FAYYAZ BAQIR

Appendices   229
Index   233

Figures

2.1 Identifying and prioritizing development operations 26
4.1 First Dialogue – Sherqila (1982) 65
4.2 RGMVP Women's Group 70
4.3 Addressing 17th Annual Session of Sustainable Development at UN General Assembly (2009) 78
6.1 Nazir Ahmad Wattoo (second from left) and Dr. Shujaat Ali, provincial secretary (third from left) launching the Bhalwal programme 104
10.1 Comparison in the joint distribution of deprivations in communities A and B 153
10.2 Deprivations in each indicator 159
10.3 Contribution of indicators to the PAMPI 160
10.4 Households' deprivation in 'X'% or more of the weighted indicators 160
10.5 Proportion of poor households across different indicators 162
10.6 Percentage contribution to PAMPI 162
11.1 Proportion of rural population with access to improved water supply in SSA 169
11.2 Proportion of urban population with access to improved water supply in SSA 170
11.3 Proportion of rural population with access to improved sanitation in SSA 170
11.4 Proportion of urban population with access to improved sanitation in SSA 171
11.5 Ratio of ODA disbursements for water and sanitation to total ODA 173
11.6 Commitments and disbursements to water and sanitation in SSA 175
11.7 Share of total gross ODA disbursement to water and sanitation in SSA (2002–2017) by project typology 176
11.8 Trend of gross ODA disbursement to water and sanitation in SSA over the period of 2002–2017, by project typology 177
11.9 Household data, SDG regions, sub–Saharan Africa, service levels 178

11.10 Household data, Democratic Republic of the Congo,
   service levels   180
11.11 Proportion of rural and urban population with access to
   improved water supply and sanitation in the DRC   180
12.1 Cereal food aid shipment and cereal production in Ethiopia
   between 1993 to 2015   195
12.2 Cereal food aid shipment as a percentage of cereal production
   in Ethiopia   195

Tables

3.1 Comparison of Bangladesh-Peru income and development
 status 52
4.1 Mandals covered in the pilot districts 67
8.1 Growth of the microfinance industry 125
10.1 Dimensions, indicators, deprivation cut-offs, and relationships
 with MDGs 156
10.2 Dimensions, indicators, and weights used for the measures 157
10.3 Programme Area Multidimensional Poverty Index (PAMPI) 158
10.4 Programme Area Multidimensional Poverty Index across
 programme regions 161
11.1 Categories of ODA allocation to water and sanitation 177

Foreword

Charting the path between the global development goals and local action has emerged as an important concern for the development community across the globe. There are numerous stories of successes and failures of the state, market, and civil society in improving global development indicators at the local level. These have generated intense debate among academics, policymakers, and practitioners. *Better Spending for Localizing Global Sustainable Development Goals: Examples from the Field* has taken an interesting approach in dealing with the issue of local action for achieving global goals, telling the stories as they unfolded on ground. These stories provide rich content for analysis, theorization, or simply replication at distant locations. One can agree or disagree with the approach taken by different practitioners on achieving their targets, but these stories invite us to think on how the approach taken may be adapted to a different context. We need to ask if achievement of global goals means a search for uniformity in action or discovering unity in diversity. In the search for underlying unity in diverse ways of dealing with local issues we find very simple and profound practices that can be taken to scale with very modest resources.

Contributors to this compendium have shown very innovative and creative ways of supporting local action. These stories show that while the provision of resources is critical in achieving results at the local level, it is the resourcefulness of local communities that plays a more important role than the size of the resources. A community's resourcefulness is exhibited in ways that challenge our imagination; for example, the case of the Southern project of a Northern NGO taking the lead in developing South-South Cooperation for poverty alleviation through supporting women self-help groups. The Southern project (an NGO) not only helps southern NGOs but the government of one of the most populous countries in the world. This happens while the governments of two countries have daggers drawn on each other. The poverty alleviation programme, which can be counted as one of the world's largest poverty alleviation programmes, succeeds where women's self-help groups (SHGs) meet only two conditions – they agree to alleviate poverty and organize. This complements Schumacher's idea of *Small is beautiful* by adding that *Small is simple, practical, and clear*. Other examples in this volume also present similar tales of brilliance, resilience, and magnificence.

We are all aware of the fact that we have reached planetary boundaries in search of an elusive goal of endless growth. One alternative way of dealing with this problem is by demonstrating cases of achieving more by spending less; and gaining more by producing less than environmentally and socially necessary. This book on better spending describes creative ways of dealing with this problem at the local level. An important example is managing scarce water resources based on the principle that "water management is water measurement." The story of a local community organization following this principle demonstrates that while water metering at the household level increases availability of water and reduces monthly water charges for low-income communities, it also increases local governments' revenues, improves governance, reverses rent capture, and leads to judicious use of water. Similarly, there is a case of developing a microfinance market in the private sector with the grant assistance of a donor community, ultimately benefitting the low-income household to gain economic independence and live with freedom and dignity.

This book seeks to explain that an alternative way of improving global targets is by spending better especially if spending more is not an immediate necessity. I would recommend this book to every academic, practitioner, and policy advocate. These stories blaze the trail of local action, and we need to come up with more such stories and collections in the days to come. The key message of the book is that "anyone can make a difference." If this message resonates with young minds, then the book has achieved its purpose.

Haoliang Xu
UN Assistant Secretary-General and
Director, UNDP Regional Bureau for the Asia-Pacific

Contributors

Shakeel Ahmad is a development economist currently working as Assistant Resident Representative and Chief of Development Policy Unit in UNDP Pakistan. Mr. Ahmad has designed and managed projects for poverty reduction, local economic development, women empowerment, and financial inclusion. He currently advises UNDP Pakistan and the government's planning institutions in Pakistan on the localization and implementation of the UN Sustainable Development Goals. Mr. Ahmad holds a Master of Science in development management from the London School of Economics and Political Science and Master of Science (honours) in agricultural economics (gold medallist) from the University of Agriculture, Peshawar, Pakistan.

Furqan Asif is an interdisciplinary environment and social scientist, with research interests broadly in human-environment relations (livelihoods, social-ecological systems, fisheries), urbanization, climate change adaptation, and natural resource management. Professionally, his experience lies within the environment-development nexus through various posts including the Canadian International Development Agency, the World Fish Center (Philippines), and United Nations University – Institute for Water, Environment, and Health. He is currently a PhD candidate in international development at the University of Ottawa and holds a Master of Environmental Science from the University of Toronto and an Bachelor of Science with honours in biology/environmental science from Western University.

Nipa Banerjee has a PhD from the University of Toronto. She has served for a cumulative period of over 40 years as a practitioner and policy analyst, and a university-level teacher in international development and foreign aid. She has worked with Canadian Universities Services Overseas and the International Development Research Center, and, for 34 years, worked in CIDA, Canada's ODA agency (now amalgamated with Global Affairs Canada). She has represented CIDA in Bangladesh, Indonesia, India/Nepal, Thailand, Cambodia, Laos, and Afghanistan, heading Canada's aid programme in the four latter countries. She joined academia in 2007.

Fayyaz Baqir is a development practitioner. His book *Poverty Alleviation and Poverty of Aid-Pakistan* was published by Routledge in August 2018. He served as the United Nation's Senior Advisor on Civil Society. At present, he is a visiting professor at the University of Ottawa. During the past three decades, he has been engaged in managing grant programmes for supporting Local Agenda 21. He has advised, researched, taught, and written on innovative practices for achieving global development goals through local action. A special focus of his work has been on poverty alleviation, participatory development, and social entrepreneurship. He has researched, taught, and lectured on innovative participatory development practices at Georgetown University, Harvard University, McGill University, Gothenburg University, Tilburg University, Wellesley College, Pacific University, University of Idaho, Punjab University, National Defence University, and Fatima Jinnah Women University.

Anthony Cholst is a highly experienced professional in international development. He has worked for 35 years at the World Bank with a geographic focus on South Asia, Central Asia, and the South Caucasus; and a functional focus on country strategy and project effectiveness. His most recent position was as World Bank Operations Advisor in Pakistan. He co-authored the 2014 book *Pakistan: The Transformative Path*. He also helped establish the International School of Economics at Tbilisi State University, Georgia (ISET). Mr. Cholst holds an MBA from the University of Chicago, an MSc from London School of Economics (LSE), and a BA from Tufts University.

Aroub Farooq is a development sector professional who works on a number of issues gripping Pakistan's economic and social development. She supports the research work of the United Nations Development Programme's Office in Pakistan, particularly in areas of multidimensional and income poverty, inequality, sustainable urbanization, and the wider implementation framework of the Sustainable Development Goals. With more than three years spent researching and working with various international development organizations and government institutions, she has developed a deep understanding of the challenges facing growth and sustainable progress in developing countries.

Shoaib Sultan Khan studied literature, law, and public administration at the Universities of Lucknow and Cambridge. He served for 25 years as a CSP officer in the government of Pakistan, 12 years with the Geneva-based Aga Khan Foundation, and 14 years with UNICEF and UNDP. Since 2005 he has been involved with the Rural Support Programmes of Pakistan on a voluntary basis as honorary Chairman and Director. Shoaib Sultan Khan is a recipient of the Magsaysay Award from the president of the Philippines; World Conservation Medal from the WWF International – president, the Duke of Edinburgh; and the highest civilian award, Nishan-e-Imtiaz, from the president of Pakistan.

Abdullah Al Mamun is a highly skilled evaluation, research, and surveys research professional with over ten years of experience in global institutions such as the Aga Khan Foundation, BRAC International, and others in Afghanistan, Bangladesh, India, Nepal, Tajikistan, the United Kingdom, and the United States. He has received academic training from world-renowned institutions such as Brandeis University (US), University of Michigan (US), University of Oxford (UK), and others. In addition, he has received several awards such as the Ontario Trillium Scholarship, World Bank Scholarship, Brandeis Merit Scholarship, Kish Fellowship (University of Michigan), and so on. At present, he is a PhD scholar at the University of Ottawa.

Adrian Murray is a doctoral candidate in the School of International Development and Global Studies at the University of Ottawa. His research explores the experiences of labour and social movements in opposition to the neoliberal restructuring of public services in the Global South. In this and other work Adrian is affiliated with the Municipal Services Project, a research project that explores alternatives to the privatization of public services in Africa, Asia, and Latin America; the Council of Canadians' Blue Planet Project; and the South African International Labour Research and Information Group. He is also an active trade unionist.

Richard Nyiawung is currently a PhD student in international development at the School of International Development and Global Studies, University of Ottawa, Canada. He holds an MA in environmental policy from Memorial University of Newfoundland, Canada (2018) and a BSc in agriculture (agronomy) from the University of Buea, Cameroon (2014). He has served as a research assistant in projects with the International Institute of Tropical Agriculture (IITA) and the Forum for Agriculture Research in Africa (FARA). His research interest is in indigenous food systems and livelihood, fisheries, development interventions and assistance.

Syed Sajjadur Rahman is a senior fellow at the School of International Development and Global Studies (SIDGS) in the University of Ottawa. He has over 25 years of experience in all facets of international development. During 1992–2013, Sajjadur worked at the Canadian International Development Agency (CIDA) (now part of Global Affairs Canada) in various senior positions including as the Executive Director of the Development Finance Innovation Working Group (2012–2013). He is a member of the leadership council at the Institute for the Study of International Development (ISID) at McGill University, Montreal, Canada.

Susan Spronk is an associate professor in the School of International Development and Global Studies at the University of Ottawa. Her research focuses on the experience of development in Latin America; specifically, the impact of neoliberalism on the transformation of the state and the rise of anti-privatization movements in the Andean region. She has published various articles and chapters on social movements and working-class formation, as well as water and sanitation politics in Latin America. She is also a research associate with the Municipal Services Project, has been a community organizer for over 20 years, and is an active trade unionist.

Neville Suh holds an MSc in agricultural economics and a BSc in agricultural economics, both from the Department of Agricultural Economics, Faculty of Agriculture and Veterinary Medicine, University of Buea, Cameroon. Since

2015, he has worked as a field supervisor with EMPOWERMENT NGO, Cameroon, focusing on agricultural extension services for resource-poor farming communities. He has also worked as a research assistant for the Association for Biodiversity, Research and Sustainable Development (ABiRSD), Cameroon. He has broad research interests in food security and agricultural research, resource management, international trade, development policies, and interventions in the Global South.

Ogochukwu Udenigwe is a PhD candidate in international development and global studies at the University of Ottawa. Her research, grounded in a deep love for humanity, currently focuses on improving maternal and child health in the Global South. She earned her master's degree from the University of Guelph in family relations and human development and bachelor's degree, with distinction, from the University of Manitoba in family social sciences. She has been involved in multiple peer-reviewed publications.

Sanni Yaya is Full Professor of Economics and Global Health, Director and Associate Dean of the School of International Development and Global Studies, University of Ottawa. His work focuses on a broad array of multidisciplinary topics in development and global health. This includes cross-cutting research and publications in disciplines of public-private partnerships, global governance, economic development, and global health. He has a strong interest in large-scale evidence (in particular, randomized trials) and analyzes of large survey data sets.

Helena Yeboah is a PhD student at the University of Ottawa pursuing the international development studies programme. She holds dual master's in applied economics and international development studies from the Ohio University, United States (2017) and obtained her BA in social sciences (economics major) at the University of Cape Coast, Ghana (2013). She has held several teaching and research assistant positions and is currently a teaching assistant at the University of Ottawa. Her research interests are in poverty reduction, economic development, and maternal and child health.

Hasanuzzaman Zaman is an adjunct researcher at the BRAC Institute of Governance and Development (BIGD), BRAC University. He started his professional research career at the Centre for Policy Dialogue (CPD) in Dhaka. In 2012, he joined the Bangladesh government's innovation lab before starting his career as an outreach manager at the Copenhagen Consensus Center. Hasan has authored peer-reviewed journals, working papers, and book chapters. His major research interests are in the areas of public administration and public sector innovation.

Part I

Citizens, state, and markets

1 Introduction and background
Global goals, local action

Fayyaz Baqir

This chapter will provide an overview of the global norm setting for supporting local action for community development, poverty alleviation, and sustainable development. Beginning with the process of local agenda-setting under Local Agenda 21 during the Earth Summit in 1992, the chapter will briefly describe the context of government-, market-, and civil society–based approaches for devolving power to local government authorities and civil society institutions to replace exclusionist administrative and fiscal practices, and overview inclusive development efforts made by civil society organizations (CSOs) and philanthropists. It will shed light on the strategies for supporting Local Agency 21 through multilateral and bilateral small grants programmes and international NGOs (INGOs) located in the formal sector. A comparative view of the formal sector initiatives and informal sector approach of community-centred initiatives will also be presented. The chapter also dwells upon the negotiating strategies, accessing mechanisms, and spending practices of the community-based programmes; especially the strategies for accessing public resources in the context of local government budgetary processes and indigenous philanthropy. The chapter will close by presenting the better spending strategies of some innovative local initiatives based on realist costing and designing for improving Millennium Development Goals (MDGs), Sustainable Development Goals (SDGs), and Human Development Indicators (HDIs) at the local level.

Keywords

Participatory development, inclusive governance, poverty alleviation, MDGs, SDGs, right-based development

Setting the development agenda: the role of market, state, and civil society

The post–World War II period was charged with the quest for equality, justice, and freedom. In former colonies this quest for freedom was perceived as fashioning former colonies in the image of their masters. During the first three decades after WWII most of the former colonies and newly developing economies followed the policy of import substitution and export-led development to achieve a high level of growth. This increased the level of national income of developing economies but did not effectively address the problem of income equality and poverty alleviation. State policies and market mechanisms both underperformed in promoting human development.

One proposed solution was to develop global goals for poverty alleviation and human development. The problem with global goals was an uneven level of development across Third World economies. Agents who were supposed to lead the process of development were not endowed with identical capacities for transforming their communities. However, there was recognition of the role of local action in achieving global goals. Recognition of local action as an important link in achievement of global goals found expression in Local Agenda 21, discussed in the Earth Summit in Rio de Janeiro in 1992.

Earth Summit 1992 at Rio rightly articulated the connection between globalization, human development, and poverty alleviation by coining the phrase "Think globally and act locally." Subsequently global thinking was spelled out in the form of Millennium Development Goals (MDGs) and Sustainable Development Goals (SDGs). Vehicles for acting locally were identified to be civil society organizations (CSOs) and local government authorities. Transfer of administrative and fiscal power to local government through devolution of power and generous grant funding to professionally managed CSOs, international NGOs (INGOs), NGOs (non-governmental organizations), and community-based organizations (CBOs) was supposed to create institutional space for anchoring global thinking in local reality. Principal authors of MDGs and SDGs had assumed that these local institutions would carry out the process of indigenization of global goals, which would culminate in human development and poverty alleviation at the local level. A quarter century down the road, the progress made in "acting locally" seems to have bogged down due to numerous barriers perhaps not anticipated by the visionaries of Local Agenda 21.

Soon after the Rio conference numerous interconnected interventions were made to realize the vision of "thinking globally and acting locally" across the countries of the Global South. Many multilateral and bilateral small grants programmes for local development were set up by Northern NGOs, bilateral and multilateral donors, and the United Nations; programmes for devolution of power to local government were introduced in many countries; and many national and international NGOs pioneered service delivery, right based, and social entrepreneurship–based programmes for human development and poverty alleviation. Success of some of the local development programmes influenced public policy and resulted in expansion of their models through government ownership. However, slow progress in improving the Human Development Indicators across the Global South points to the need for in-depth analysis of the perils and promises of the process of local institution building for human development. In operational terms, this means identifying the indicators of institutional maturity and barriers in the way of ascendance of local institutions to play their due role – specifically, accessing public resources for human development and poverty alleviation – and finding effective remedies for crossing the barriers based on the case studies of selected interventions. This study aims at undertaking the analysis of the dynamic link between local institution building and achieving globally inspired targets through the presentation of selected case studies and frameworks tested in the field.

Financial underperformance at the local level

A close review of Southern economies shows that weak human development goes hand-in-hand with underutilized financial resources in the public sector. The resources allocated and ground conceded by the centralized governance structure to local government authorities and civil society institutions have not led to improvement of human development indicators, especially in the case of low-income communities. Extreme poverty and low levels of human development indicators cannot be explained by the low level of budgetary allocations. What needs to be explained is how existing levels of domestic and external resources do not produce the desired result. Some donors have tried to camouflage the issue of underperformance by using the amount of funds spent as a proxy for "result-based management for measuring their success"; others have presented their failed projects as "best practices" in international fora.

Defining the failure in terms of underperformance rather than underfunding or insufficient legislation provides a very powerful perspective for dealing with the issues of poverty alleviation and human development at the local level. Experience of many low-performing economies clearly shows that failure in achieving human and social development goals is caused by continuing tension between representation-based "global" policy instruments – inspired by global norms – and participation-based local institutional practices – required for local development. The cases presented in this book will review human development practices in relation to the dynamic conflict between the representative and participatory approaches of citizen engagement for optimal realization of human development potential. Specifically, we will explore why the resources allocated for poverty alleviation and human development do not convert into desired outcomes at the local level.

Cases presented here will examine the link between the "global" goals and 'local' action in view of slow human development in the context of asymmetry of powers, capacities, and structures. The postcolonial societies have not been able to achieve the level of human development commensurate with the level of resources made available for this purpose. It is important to note here that it is not only the official development assistance, but also private investments, philanthropic donations, remittances sent home by immigrant workers, and assets created by the competitive and resilient informal sector that have created the pool of resources for low-income communities in developing countries. Despite the availability of resources from all these sources provision of basic services is not sustainable and service delivery systems are not fully functional. This calls for an in-depth analysis of the delivery capacity of the local government authorities and CSOs; identification of the gaps created by institutions representing but not engaging the poor in local development; and the enabling environment needed to change this practice. This book will build on the case studies of practices, processes, and strategies used for facilitating community access to public resources for human development.

Civil society and local development

Professionally led CSOs entered human and social development sector in the Global South in the 1980s as service providers; they soon realized that without engaging and activating the government they could not achieve a significant scale, and at the beginning of new millennium they made a transition to a right based approach. Right based work has not had much success. Realizing the limitations of a right based approach, some CSOs tested social business approaches based on the concept of "affordable user fee, cross-subsidies, and targeted marketing." This approach has led to local success stories which need to be documented, analyzed, and presented as business models for replication through self-selection. Such models show the way for aid effectiveness, better spending, and sustainable human and social development. These models exist in health, education, income generation, and human development sectors. The present research will present some of these models.

It is also important in this context to review inclusive governance, MDGs, and other indicators of poverty alleviation and social change as a relationship between the states and citizens. This calls for understanding the nature of existing relationships and the means being employed to change these relationships to achieve the global goals. In the first place, it is important to note that the process of drafting most of the global conventions, treaties, goals, and frameworks has been carried out in an institutional arrangement where no validation and feedback system exists to check for social relevance and effectiveness. Second, articulation of these agreements assumes that the agreements are converted into national legislation by signatory states and assumes the existence of an institutional mechanism to seek compliance with these commitments and legal instruments.

Third, a very big assumption is being made that parties to these agreements understand the reality of the poor at least as much or better than the poor understand it themselves. Fourth, it is also assumed that in the absence of any accountability clauses, the states responsible for abiding by these laws and development goals would comply with the obligations they have agreed to endorse. Finally, no validation and indigenization actions have been required from the signatory parties to align the provisions of these rights and goals with the social conditions of the poor people based on the feedback received from them.

The primary difference between the global goals inspired by Western development norms and local reality has been summed up by 1992 Nobel Peace Prize winner Rigoberta Menchú as the difference between "the organizing principle of the market" and "organizing principle of the community" (Groucher, 2004, 24). In Menchú's words, "We see the community, not the market as the building block of the modern self-reliant development based on cooperative village life" (ibid.).

Due to the process of rapid globalization, urbanization, and capitalist development, the nature of the community economy has changed, but the principle of the community economy still governs the lives of 80% of the world population. Third World poor, who constitute these 80%, live under what is now called an informal economy (De Soto, 2000; Hasan, 2007). However, development of local models has been heavily influenced by the outlook of formal sector professionals. During the post-WWII period poverty has been investigated from the point of view of the professionals. Measures of poverty designed by these professionals have included indicators based on income, purchasing power parity, calorie consumption, and access to services and opportunities. These measures have one thing in common: they have been designed by the professionals not by the poor themselves. If we take into consideration the question, "Whose reality counts?" (Chambers, 1997), we need to find out how the poor look at the poverty themselves. Would consulting the poor make any difference in terms of finding the solutions? If so, how do we diagnose and alleviate poverty in consultation with the poor themselves. In this connection some pioneering work done by community practitioners has brought to light some very critical findings.

Three chapters in this volume deal with the effectiveness of professionally developed frameworks in the context of citizen-state relationships. First, there's **Anthony Cholst's** narrative of **An anatomy of a loan.** Then there are two stories on the use of well-developed frameworks for facilitating better spending. One experiment was conducted in Afghanistan on use of **Multidimensional poverty measurement and aid efficiency.** Written by **Abdullah Al Mamun and Sanni Yaya**, the case presents the **use of a multidimensional poverty assessment (MPA) approach** to increase the aid efficiency of poverty reduction programmes in Afghanistan. The Multidimensional Poverty Index (MPI) has been adopted by several countries to develop better anti-poverty policies and to increase the efficiency of financial expenditure. This approach is in the experimental phase and would be applicable at the institutional level as an effective and efficient tool to evaluate the impact of aid in poverty reduction programmes in meeting Sustainable Development Goals (SDGs). The second case is narrated by **Hasanuzzaman Zaman and Syed Sajjadur Rahman**, who, in their chapter on **Improving SDG spending through South-South SDG tracking,** describe how South-South Cooperation focused on knowledge sharing for tracking SDG implementation can help improve SDG spending. The case provides insights about how the use of the SDG tracker by the Finance Division and Planning Commission of Bangladesh for budgetary decisions set the scope for analyzing a policy transfer between Least Developed Countries (LDC) (Bangladesh) and a middle-income country (Peru). It illuminates new ways for improving SDG spending through existing administrative arrangements.

Dimensions of the social space for achieving development goals

Since the state plays an important role as promoter of global goals and provider of financial resources for local development, it is important to review the space defining state-civil society interaction. As described by Professor Adil Najam, CSOs have four distinct options in their interaction with the state; One, they may have similar goals and similar means, which would lead to a cooperative relationship. Two, CSOs may have similar goals and dissimilar means resulting in complementarities. Complementarity would mean that citizens and the state could have a very fruitful collaboration. For example, both the CSOs and the state agree on poverty eradication as their goal. The state has the budgetary resources, legislative powers, and administrative authority at its disposal and citizens might have community outreach, organized interest groups, and technical and social knowledge, which, combined with the resources of the state, could produce extremely positive outcome. Three, the state and CSOs might have dissimilar goals but similar means, in which case engagement is possible only through co-optation. This will lead to a situation where compromise is only possible by surrendering the claims of the communities. This situation would call for a long and drawn out advocacy campaign based on evidence, dialogue, and public pressure. Four, in a situation where CSOs and the state have dissimilar goals and dissimilar means, claiming the rights would lead to confrontation.

CSOs would have to clearly distinguish between all four options and devise strategies accordingly (Najam, 2000). This means that just situating local action in civil society would not provide sufficient conditions for dealing with the market and policy failure. It is also important to mention here that the lower tier of civil society consisting of community-based organizations cannot perform effectively without a higher tier functioning as a support organization. Some of the chapters presented in this book will highlight the role of the support organization resulting in effective performance of CSOs. One example is the role of the Aga Khan Rural Support Programme (AKRSP). AKRSP's guidance and support led to doubling the income of rural families in its programme area in ten years. Subsequently AKRSP's general manager introduced this approached to the government and NGOs in India, lifting millions of low-income households through the formation and support of women's self-help programmes.

Social capital and global goals

Civil society intervention is seen as a panacea for addressing shortcomings in human development, dispensing economic justice, alleviating poverty, mitigating climate change, and ensuring sustainable development in view of market and public policy failures. Civil society's secret to success has been presented in various names: participatory development, right based discourse, trickle-down and bottom up approaches, social entrepreneurship, voluntary economy, social capital, micro-initiatives, and self-help groups, to mention a few. It is asserted that civil society succeeds where market and government fail. In my view the government and market constitute the professional constituency of knowledge, and civil society seems to have an edge in tapping the tacit knowledge of informal communities. The formal economy represents the global, and the informal economy personifies the local.

Success in social capital formation in this context can be defined as the development of a shared knowledge space – a prominent form of social capital – carved out for strengthening interactions between the formal sector professionals and informal sector citizens. What makes civil society succeed has various explanations: micro scale, below market costs, responsiveness to local priorities, economies of scale, social accountability, and community economics. What counts is the lead role taken by local communities and facilitation by the external experts. We need to acknowledge the fact that the state, private sector, and civil society all have their share of successes and failures in reaching out to the underprivileged, and success is defined by their modes of interaction with the beneficiaries. There are many Ramon Magsaysay Award recipients in the public sector who have been recognized for their inclusive development work across the Asia-Pacific region. There are also numerous philanthropists, social entrepreneurs, and civil society initiative developers blazing the trail of local development. An illuminating example is provided by the support work carried out by a community professional in restoring a dysfunctional water supply scheme in a small town in Pakistan and establishing a sustainable and affordable water supply system based on the principle of "water management is water measurement." This model helped the poor by developing a "business model" for providing water to metered connections. This model taught the government to operate like a private sector organization, encouraged community members to demand quality services as consumers, and produced benefits characteristic of a public sector organizations.

A deeper probe into the nature of the relationship between the state, civil society, and disadvantaged communities reveals that social capital formation is a three-dimensional activity. The first dimension is the sector in which a development initiative originates – that is, state, market, and civil society (also known as first, second, and third sectors). The second dimension is the track followed to connect with the communities. This track unfolds along one of the three dimensions: power (legislation and rule-making), finance (fiscal and budgetary provisions), and people (participating in decision-making) – or first, second, and third vectors. The third dimension is the space in which the interaction unfolds: cooperation, confrontation, and collaboration – named here as first, second, and third spaces (Najam, 2000; Boris & Steuerle, 2016). My contention is that only the location of an enterprise in the third sector does not guarantee successful social capital formation. It is not location but participation that counts.

A critical evaluation of the experiences of participatory initiatives across the Global South reveals that it is not the origin of an activity in civil society, but the path followed and the space selected for engagement that determines the accumulation of social capital. In this process it is not the lead taken by civil society but the process of civic engagement that plays a critical role. Access to opportunities for social capital formation is facilitated by the discourse based on engagement and collaboration. This assertion calls for a critical appraisal of the prevalent notion of civil society, state, market, and communities in relation to trust formation. A critical appraisal of the insights from the field provides a clue to the conditions that are necessary and sufficient for social capital formation. It is social capital that forges the dynamic link between the global and local. An interesting example is provided by the evolution of the microfinance sector in Pakistan. The funding for developing the market and enhancing the capacity of CSOs was provided through grants and loans allocated through official development assistance; delivery of service was carried out by CSOs working as private sector organizations; and benefits were generated for low-income communities. The distinction of state, market, and civil society collapsed in the sector as all the actors embraced participatory approaches and abandoned their conventional roles.

The connection between global goals and local action presupposes the key role played by the state in converting financial inputs into social outputs. Our misplaced expectation from state as the key actor in development industry is caused due to our misunderstanding of the strengths and limitations of the postcolonial state. There is a range of views on state. The classical Marxian framework looks at the state as an instrument of the exploitation and oppression of one class by another. In this view underperformance is built in the nature of the state. Postcolonial theorists view the postcolonial state as an overdeveloped structure in relation to underdeveloped societies. Pioneering work by some development practitioners and academics pointed out that the postcolonial states inherited developed law and order administrations but lacked a development administration depicting an undeveloped state with weak technical and enforcement capacity and limited capability for realizing the "trickle down" effect. Economic experts argue that postcolonial states lacked resources to realize their own development objectives (Haq, 1966; Sachs, 2005). That is why Swedish economist Gunnar Myrdal called the Pakistani state a soft state, something having a strong hand but delivering a soft punch. Such a soft state is different in character from the nation state in the Global North.

Due to the extremely limited capacity of the "soft" state to realize the goal of local development in former colonies, civil society has emerged as the most important actor for poverty alleviation and local development. Civil society is engaged with the state in various capacities: as a partner, a contractor, or a confronting force. It can act both as facilitator or "representative" of communities, and in the process it either enhances the range of their choices or reproduces existing relations. This book looks at the matrix of state, civil society, market, and the community relations with reference to creation of access to development opportunities for low-income communities and social capital formation. An analysis of the gaps between the prevalent perceptions and ground reality of various actors and sectors provides the clue to the nature of engagement that leads to social capital formation. **Nipa Banerjee** has given an insightful account of the very inspiring work of the Bangladesh Rural Action Committee (BRAC). BRAC is one of the driving forces behind Bangladesh's graduation to a middle-income country and why the country is reaching most of the global development targets included in MDGs. BRAC's interventions have contributed to large-scale, positive changes through the implementation of cost-effective community-level economic and social programmes that enable men and women to realize their potentials and use their assets to come out of poverty. It is an illuminating story of partnership with the poor and the disenfranchised and of responding to local needs for sustained impact on poverty reduction.

Banerjee's account of **Afghanistan's National Solidarity Programme** implemented in a fragile situation is a surprise development success story emerging out of Afghanistan, one of the world's poorest countries and the deadliest for terrorism. The NSP was conceived in 2002 to give Afghan villages a democratic local administration, along with access to basic services. It subsequently developed the ability of rural communities to identify, plan, manage, monitor, and maintain their development projects, making the development results sustainable. The NSP was conceived, cultivated, and developed according to Afghan traditions and values of unity, equity, and justice. The empowered rural communities followed the global goal of leaving no one behind. Chapter 2 has provided a hypothetical case depicting the common perceptions of donors, states, civil societies, markets, and communities and options available for making astute choices based on a good understanding of the process of better spending in the public sector.

This point can be better explained with reference to Pakistan. A close review of Pakistan's economy shows that weak human development goes hand in hand with Government's under performance. This underperformance is reflected in various ways. One, there is underutilization of financial resources in the public sector (Baqir, 2007). Two, Government's capacity to mobilize resources is also limited. Existence of a large philanthropic sector- double the size of public sector's social development budget clearly shows under performance in mobilizing resources from citizens (AKDN Pakistan, 2000). We see a large philanthropic sector side by side with a narrow taxation base and a large size informal sector with significant asset base indicating the effectiveness of philanthropists and low-income classes in meeting societal needs (Rahman, 2013; Hasan, 2006). We need to note two dimensions of the problem of better spending in this regard: one, the resources allocated, and legal and fiscal space conceded by the state for provision of basic services to the low-income communities, have to a large extent not been accessed by the civil society for full realization of human development potential. Basic services have been received by the communities through what has been termed as "uncivil society," anchored in the informal sector. Two, the informal sector in large swathes of the Asian, African, and Latin American economies has a sizeable asset base, but its assets are not integrated with the mainstream economy (De Soto, 2000; Anzorena, 1998; Mustafa, 2005; Sattar, 2011; Hasan, 2006). This is not only the state's failure but the civil society's failure as well. This requires a deeper view of the missing link at the local level.

Dealing with the missing support system at the local level

Eminent South Asian development practitioner Akhter Hameed Khan pointed out that three essential infrastructures were needed for local development: administrative, political, and social. All these infrastructures did not exist below the district level in most of the former colonies at the time of their independence. Khan very insightfully discovered that the collapse of the feudal authority created a deficit of social capital, which contributed to a lack of human development. This called for embarking upon the process of human development by creating appropriate institutions and designing appropriate methods to fill the void created by the collapse of traditional institutions. He realized that the existing jurisdiction of the police station could be converted into a viable unit for administration – being responsible for (i) the provision of services and (ii) up-gradation of skills (iii) outreach to all the low-income settlements and households. The state's underperformance can be reversed by following innovative approaches for creating new social infrastructure based on community interests instead of community, clan or tribal loyalties. Specifically, these innovative approaches have consisted of four concrete steps: (i) initiating social organization, (ii) undertaking documentation, (iii) invoking standard operating procedures of the government to implement projects identified by communities, and (iv) conducting dialogue to improve delivery. He believed that filling the financial gap without filling the institutional gap led to mismanagement of resources rather than the efficient use of resources – efficient use that would enable the country to find solutions for economic challenges "within the existing system and with available means."

It is important to note here that CSOs don't represent a homogenous world. There is a wide range and many levels of professionalism and a diverse resource base of CSOs across the globe (UN, 2004). CSOs include professional interest groups, trade unions, resistance movements, charities, social welfare, and service delivery organizations, as well as right based networks and donor-sponsored contractors. Their objectives may be as varied as subsidizing service delivery, investing in local institution building, or providing credit to micro-enterprises; poverty alleviation, agitation and negotiations for claiming basic rights, or serving as watchdogs of public institutions. The basic characteristic which defines the difference between a cop-out CSO – maintaining the status quo – and a generative organization that harnesses and unleashes the potential of local communities to realize higher level of self-development is the means employed to achieve the intended goals. These means fall into two broad categories: following a template produced in a context not related to the reality of poor or initiating a process to help communities identify their priorities and develop their capacity to achieve them with their own efforts. The so-called right based rhetoric and poverty alleviation jargon can be very deceptive in equating the contractors implementing the template with the facilitators helping communities realize their potential. Similarly, use of the civil society label does not make an initiative a self-help project and the affiliation with government a denial of an inclusive framework. The outcome of a public, private, or voluntary institution's intervention will depend on the support system it puts in place. The extent to which CSOs succeed in seeking access to state and market resources is not hampered due to power or resource deficit of CSOs but due to their undeveloped capacity to overcome the trust deficit.

Three cases are important to note in this regard. Furqan Asif writing on **Aid effectiveness** tells the story of a citizen watch group, the Urban Resource Centre (URC), in Pakistan's metropolitan city Karachi. URC used a unique strategy of integrating citizens' tacit knowledge in the process of official development planning and implementation to make the planning and implementation process responsive to citizens' needs and aligned the budgeting process with the social conditions of the city's poor, paving the way for better spending and judicial use of public resources.

Sanni Yaya, Ogochukwu Udenigwe, and Helena Yeboah have documented the case of **Development aid and access to water and sanitation in sub-Saharan Africa (SSA)**. They expand on the links between Official Development Assistance (ODA) and access to safe drinking water and sanitation in SSA while examining the impacts of inefficient water supply models. Making the case for moving **from more spending to better spending for food and nutrition security in Ethiopia**, Sanni Yaya and Richard Nyiawung argue that for dealing with conflict and natural disaster and improving food security two developmental frameworks adopted by the Ethiopians have played a very effective role. The advancement of agricultural technologies and infrastructures through direct income investment as well as the advocacy of viable interventions to promote growth amongst the poorer populations have led to scalable activities for hunger relief.

Adrian Murray and Susan Spronk have made the case for **alternatives to privatization for the improvement of water and sanitation services** in Cape Town, South Africa, and Montevideo, Uruguay. They depict how working in partnership with social organizations, managers in public utilities improved water and sanitation services. It is interesting to note that a similar experiment succeeded in improving water supply services in an urban settlement in Pakistan as well. While the public managers in Cape Town and Montevideo promoted principles of "social efficiency" rather than the narrow notions of "financial efficiency," partners in Pakistan developed a social business model to promote more equitable outcomes in service delivery.

Local agency: participation or representation?

The ineffectiveness of the government, private sector, and CSOs arises in dealing with the universe of the unwritten, unrecorded, and undocumented social reality of low-income communities and marginalized groups; their living conditions; their human, social, and financial assets; and the condition of the existing social, administrative, and physical infrastructure in their settlements. This unrecorded world can be transformed in the process of observing, recoding, and transforming the existing practices with the participation of the people based on redefined roles and responsibilities. This world can be changed not through new templates but through new processes. Templates reproduce existing relationships. The process of accessing, creating, and possessing documents empowers communities to deal with the inflexibility of templates and their incompatibility with reality on the ground.

Social change is a process that leads to the transformation of economic, social, and political relations of power that protect the rights of the citizens. It is important here to make a distinction between CSOs that *alleviate* the detrimental effects of existing relations of power by working within the system and filling the gap due to the limited reach of the government for service provisions and the ones that want to *alter* these relations. Altering the relationship with the state is not possible through local action, but alleviation of detrimental effects is. When CSOs fail to alleviate the detrimental effects of exclusionist practices of the state institutions, "un-civil society" takes over. This happens when CSOs dealing with government and market failures do not create *designs aligned with the reality of* poor. This can be done through what has been termed "social preparation" by Arif Hasan (Hasan et al., 2015).

Formal-sector experts see the failure of CSOs as the failure of communities. Differences in the living conditions and strategies of survival of these communities is not seen as an expression of diversity but as a lack of awareness, assets, finances, and power of the people. Based on this vision, they consider it "their burden" to provide handouts and subsidies to the poor and guide them to seek their rights. They don't consider it important to gain an understanding of the people living in poverty and suffering from discrimination. They look at people's lack of formal knowledge as illiteracy or ignorance (Chambers, 1997).

They don't even know that low-income communities have successfully established their own water, sanitation, shelter, health, education, and job-creation systems. They meet their daily needs with their own resources and manage to survive under very adverse conditions. Limited capacity of the formal sector to help the poor can be gauged by looking at the size of the unspent social sector budget of the government, number of ghost facilities in the education and health sectors, and the magnitude of unutilized or misappropriated funds in the public sector. The formal sector does not even have maps of most of the settlements and no documentation of the informal economy exists at all. They overlook the fact that the size of indigenous philanthropy and remittances of overseas workers is many times the size of the official development assistance (ODA) flowing to the countries of the South. Dispensation of justice to the majority of rural communities still takes place through informal justice systems with the full cooperation of local authorities (AKDN Pakistan, 2000; AFP, 2019; UNDP, UNICEF, UN WOMEN, 2012).

They have no explanation for the fact that while people are willing to pay tribute, extortion money, and security charges to local mafias, they do not pay taxes to the governments. They cannot explain why the poor prefer regressive informal practices over rational formal practices. There are theoretical and practical models available for mobilizing communities to negotiate their rights, but most of the donor-driven CSOs seldom go beyond project completion. Praxis models – integrating word with action – demonstrate the path for moving from the perception of disenfranchised people as people with "deficiencies" to action rooted in their strengths and capacities (Bhatti, 2015). These models include the diagnostic dialogue method of Akhter Hamid Khan (Khan, 1996) based on the principle of working within the system and living within the means, and Paulo Friere's method of challenging the power structure (Millard, 1986). These models build on the agency of the disenfranchised rather than their representation. What distinguishes actions upholding freedom from restriction of the marginalized communities is whether these actions strengthen the agency of these communities or representation of their aspirations by the professional elites.

Finally, an interesting case to mention in this respect is that of indigenous peoples here in Canada. First Nations in Canada can rightly be considered a part of the Global South situated in the North. In this volume we had hoped to include experiences of promoting the well-being of First Nations in Canada through **interventions based on the social reality of indigenous peoples**. However, after an extensive search we were unable to find an appropriate contributor. This failure highlights the deplorable state of the relations between indigenous peoples and the – state in Canada and the urgency of meeting the needs of indigenous peoples.

References

AFP. (2019, April 8) *Remittances to Developing World Hit Record in 2019*. World Bank. https://www.france24.com.

AKDN Pakistan. (2000) *Philanthropy in Pakistan, AKDN Report*. Islamabad, AKDN.

Anzorena, J., Bolnick, J., Boonyabancha, S., Cabannes, Y., Hardoy, A., Hasan, A., Levy, C., Mitlin, D., Murphy, D., Patel, S., & Saborido, M. (1998) Reducing urban poverty: Some lessons from experience, *Environment and Urbanization*, 10(1), pp. 167–186.

Baqir, F. (2007) *UN Reforms and Civil Society Engagement*. Islamabad, UNRCO.

Bhatti, R. F. (2015) *Exploring Strategies for Effective Advocacy: The Lived Experience of Leaders of Pakistani Non-Governmental Organizations*. PhD dissertation. San Diego, CA, University of San Diego.

Boris, E. T., & Steuerle, C. E. (2016) *Nonprofits & Government: Collaboration & Conflict*. Washington, DC, The Urban Institute.

Chambers, R. (1997) *Whose Reality Counts? Putting the First Last*. London, Intermediate Technology Publications.

De Soto, H. (2000) *The Mystery of Capital: Why Capitalism Triumphs in the West and Fails Everywhere Else*. New York, Basic Books.

Groucher, S. L. (2004) *Globalization and Belonging: The Politics of Identity in a Changing World*. Oxford, Rowman and Littlefield.

Haq, M. U. (1966) *The Strategy of Economic Planning: A Case Study of Pakistan*. Karachi, Oxford University Press.

Hasan, A. (2006) Orangi pilot project: The expansion of work beyond Orangi and the mapping of informal settlements and infrastructure, *Environment and Urbanization*, 18(2), pp. 451–480.

Hasan, A. (2007) The urban resource centre, Karachi, *Environment and Urbanization*, 19(1), pp. 275–292.

Hasan, A., Ahmed, N., Raza, M., Sadiq, A., Ahmed, S. U., & Sarwar, M. B. (2015) *Karachi: The Land Issue*. Karachi, Oxford University Press.

Khan, A. H. (1996) *Orangi Pilot Project, Reminiscences and Reflections*. Karachi, Oxford University Press.

Millard, E. J. (1986) *The Investigation of Generative Themes in ESL Needs Assessment*. Master's Thesis. Vancouver, BC, University of British Columbia.

Mustafa, D. (2005) (Anti)Social capital in the production of an (UN)Civil society in Pakistan, *Geographical Review*, 95(3), pp. 328–347.

Myrdal, G. (1968) *Asian Drama – An Inquiry into the Poverty of Nations, Volume 1*. New York, The Twentieth Century Fund.

Najam, A. (2000) The four C's of government third sector-government relations, *Nonprofit Management and Leadership*, 10(4), pp. 375–396.

Rahman, P. (2013) *We Are All Ninja Turtles of Mapping*. Talk at Asian Coalition of Housing Rights' (AHCR) Gathering in Bangkok. Bangkok, ACHR.

Sachs, J. (2005) *The End of Poverty: Economic Possibilities for Our Time*. London, Penguin Press.

Sattar, N. (2011) Has civil society failed in Pakistan? *Social Policy and Development Centre (SPDC)*. Working Paper 6. Karachi, Pakistan, SPDC.

UN. (2004) We the peoples: Civil society, the United Nations and Global Governance, *UN Report of the Panel of Eminent Persons on United Nations – Civil Society Relations*. New York, United Nations.

UNDP, UNICEF, UN WOMEN. (2012) *Informal Justice Systems: Charting a Course for Human Rights Based Engagement*. New York.

Part II

Frameworks for better spending

2 Anatomy of an effective development operation

A finance minister considers whether to borrow from the World Bank[1]

Anthony Cholst[2]

International Financial Institutions (IFIs) are often criticized for lending large sums to developing countries, thus adding to countries' debt levels. At the same time, the needs of developing countries are real and large, and such financing represents a potentially important source for meeting these needs. How can countries decide whether, and under what circumstances, to take such financing? This chapter takes the point of view of a Minister of Finance who is trying to decide whether to take a World Bank operation, and guides him through all of the key elements of the operations "package" including the different lending choices available – investment, policy-based, results-based, guarantees, technical assistance – as well as other operational elements impacting effectiveness – from prioritization, design, and cost-benefit through implementation issues including staffing, procurement, auditing, and monitoring. For each element, the author, based on his experiences, provides examples and views on the principles and requirements which will determine whether the operation will achieve sustainable development. It highlights that the two most important factors in development effectiveness throughout the process are ownership at the country and community level, and a laser-like focus on results. The goal is neither to glorify nor demonize IFI lending but rather to give the reader the tools to evaluate for themselves when such lending is likely to be effective or not.

Keywords

International financial institutions, World Bank, loans, country ownership, development effectiveness, cost-benefit, sustainable development, debt

Introduction

The finance minister of a medium-sized developing country was having a coffee with a parliamentarian on a hot sunny afternoon and was feeling very apologetic. The parliamentarian was asking for all sorts of support for his constituents, none of which the minister had the budget to fund. But the parliamentarian had an idea – if the budget could not support the activities, perhaps an international financial institution like the World Bank could. The finance minister was ready to consider this but was also cautious. He knew that the needs of the citizens were real but that there was more than money involved in taking World Bank funds to support an operation.

In order to make such a decision, the finance minister decided to take stock of all of the things that come with taking a World Bank–financed operation – in other words, the World Bank "package." This includes understanding (i) the basic principles and parameters under which the World Bank functions, (ii) how the World Bank selects which areas to support, (iii) how World Bank operations are designed, (iv) types and terms of financing, (v) how staffing works in World Bank operations, (v) how funds would flow during implementation, and (vi) the process of project monitoring and evaluation. And then within each of these elements, the minister would try to understand what factors determine whether the operation is likely to be more or less effective.

Basic principles and parameters of the World Bank

Back in the office, the finance minister requested that an aide provide a brief on the basics of how the World Bank works. Luckily, the aide had some experience working with international organizations, and was happy to show off his knowledge of the World Bank's basic approach:

The World Bank is jointly owned by both developed and developing countries, and unlike a private company, profits are not the objective. It was first established for the purpose of helping to reconstruct war-torn economies after WWII, but over time it adopted the further goal of helping to build a world without poverty, which more recently expanded to the twin goals of eradicating extreme poverty and promoting shared prosperity.[3]

These are admirable goals, pursued by over 10,000 dedicated and highly trained staff, recruited from around the world. And remarkable progress has been achieved towards these goals – extreme poverty has dropped from 36% in 1990 to 10% in 2015. Since then, however, progress has slowed, particularly in Africa, and levels of poverty reduction vary greatly between countries. Just as importantly, there are many factors at work beyond the World Bank in influencing the level of global poverty.

A second key parameter is the level of finance it provides. In fiscal 2018, the World Bank committed US$47 billion to over 120 countries, making it the largest and arguably the most influential of international financial institutions (World Bank Annual Report, 2018b). For some of the poorest countries, international development financing can amount to a significant part of their economies.[4] And even in countries which have access to alternative official or private external sources, the incremental addition of resources can make a huge difference.

However, it is equally true that developing countries have access to more financial sources than in the past. The Asian Development Bank provided US$20 billion and the African Development Bank provided US$8.7 billion in 2018 (ADB Annual Report, 2018; AfDB Annual Report, 2018)[5] Other large international financial institutions include the European Development Bank (EBRD), the Inter-American Development Bank (IDB), the Asian Infrastructure Investment Bank (AIIB), and the New Development Bank (NDB). Private international investment flows to developing countries are even larger than official flows and growing, although they are concentrated in the wealthier developing countries.

The World Bank has three principle delivery mechanisms – financing operations, creating and spreading knowledge (e.g., preparing technical analyses), and convening power (e.g., conferences, ranking tables comparing countries' progress against different measures such as doing business and human development, talks focusing attention on critical issues of the day, the annual World Development Report). The World Bank is able to be most effective when these three mechanisms are used in a coordinated manner to meet the demands of developing countries.

As regards operations, not all World Bank financing is provided as loans. Support to the poorest and most conflict-affected countries is provided as grants (for example, for Afghanistan). However, the majority of development financing provided by the World Bank and other international financial institutions is provided as loans. Of the US$47 billion provided by the World Bank in fiscal 2018, 49% were market-based loans and guarantees, 40% was concessional loans and guarantees, and 11% was grants.[6]

Frankly, the finance minister was not very impressed with the World Bank's overall goals and global statistics. What he really wanted to know is why the country should seek World Bank financing for a specific operation. That is, what makes the World Bank different from other sources of finance? The aide was a bit flustered but attempted to answer the question:

The World Bank stands out from other international financiers in its global, sector, and country reach. It is active on all the regions of the world, and this gives it the ability to compare and contrast experiences from across the world. Its activities span all the main development sectors – human development, infrastructure, governance, the environment, agriculture, water, rural and urban development, and so on. This gives it a very rich array of expertise to offer. Finally, it has significantly decentralized, with well-developed and staffed offices in more than 120 countries. More than half of its operational staff are now, in fact, in the field, including nationally hired sector experts. This gives it country depth to match its global and sector knowledge.

Another key distinguishing feature of World Bank operations is that they are generally provided on-budget and through the government. This is in contrast to some donors or international agencies that provide grants off-budget directly to non-governmental organizations (NGOs). Some donors have developed a strong relationship with a particular NGO or may feel that going around the government is necessary in situations of extreme corruption or a weak state.

However, what off-budget operations may gain in service delivery is counterbalanced by reduced country ownership. Going through the budget also provides transparency in funding, builds country capacity, enables scalability, and encourages sustainability. And where there is limited state capacity, World Bank operations can be channelled through the government's budget to NGOs to provide services, but within the overall government framework.[7]

The finance minister wondered if the aide was overly enamoured with the World Bank. While he was happy to learn about these positive qualities, he also knew that some scholars and think tanks have criticized the World Bank. He asked the aide to be more frank and give the weak points as well. The aide was now more flustered than ever, but responded that most criticisms fall into three groups:

The first set of criticisms points to the fact that a World Bank loan will inevitably increase a country's debt level, and if the loan doesn't result in increased income, it can place an unsustainable burden on the country. The World Bank also has institutional pressures to lend more, beyond the impact of a particular operation. Therefore, it is critical that the government itself carefully considers the project's economic and financial returns, and lower cost or grant alternatives if available. This being said, World Bank operations are provided at very competitive market and sometimes below market rates, and borrowing, if done well, can increase growth.

A second set of criticisms revolves around the underlying policies that influence and shape World Bank operations. Some critics feel the World Bank often offers boilerplate free-market solutions that may work in some countries or regions, but do not take into account specific country conditions or local sensitivities or views. Other critics feel that some of its operations do not adequately take into consideration environmental and social costs. In response, the World Bank has adopted environmental and social safeguard policies, and requires stakeholder consultation and mechanisms of redress. Still other critics are concerned that because of these efforts, its operations have become highly cumbersome, adding to the time and cost of taking its funds. There is, thus, a trade-off between addressing these policy concerns and the resulting additions to the complexity of operations that needs to be considered.

Finally, there is a third set of concerns that the benefits of operations can be captured by both local elites and multinational corporate interests, either through corruption or system-wide structural causes. As noted, World Bank operations are predicated on assisting the poor and the growth of the middle class, and have fiduciary controls to prevent corruption, but leakages can occur. At a global level, the drop in extreme poverty has indeed been accompanied by a worrisome increase in global income inequality. At an institutional level, one study of 600 project completion reports found the potential for possible corruption or elite capture in about 20% of projects, though these references are rarely straightforward and more often refer to "irregularities" in procurement or financial management (Winters, 2010).

Indirect impacts also need to be examined. For example, energy projects often aim to help reduce power costs, but they can result in a greater benefit to the elite who use more power unless there are offsetting actions. Therefore, it is critical that direct and indirect benefits be evaluated for their potential for elite capture and their focus on the poor and middle class. This being said, it is difficult to fully ring-fence operations against elite capture.

During the tenure of President Wolfensohn (1995–2005), the World Bank took a number of actions to respond to these criticisms, including moving more staff to the field, increasing the focus on country ownership, providing some debt relief including shifting to grants for the poorest countries, and highlighting the challenge of corruption. More recently, the World Bank has made its data and analysis more open to the public and has brought more attention to global challenges including communicable diseases and nutrition, renewable energy, citizen engagement, fragile and conflict-affected areas, and refugees. These changes have brought the World Bank closer to the client and deepened its technical reach. Nevertheless, its tendency to look overly inward for knowledge and solutions continues to persist, as does its strong built-in pressures to lend.

The finance minister thought over the briefing from his aide. It was clear that the World Bank in many cases has the financial strength and knowledge reach to support a wide range of development operations. It was helpful to know that in most countries its national presence and awareness of local approaches has grown and that it provides support primarily through the government's own budget. At the same time, it was important to be aware of World Bank lending pressures, the propensity to look inward for knowledge, and the potential for elite capture. It would be important for the government to do its own due diligence on these issues, including evaluating economic, environment, and social costs, with stakeholder consultations, before moving forward.

Identifying the right operation

The finance minister drummed his fingers on his desk. He was ready to cautiously explore a World Bank financed operation, but the parliamentarian had presented a wide range of needs. He wanted to narrow them down and only seek support for the highest priorities with the strongest payoffs. He knew that even grant financing had a cost – the opportunity cost of not doing alternative activities – which required selectivity.

Picking the right operation would therefore be essential. His aide had informed him that the World Bank's review of results and performance found that only 73% of its projects that closed from FY14–FY16 were rated moderately satisfactory or above (IEG, World Bank, 2018b). He hoped the World Bank could help him with making a decision based on its experience and felt that it was time to call in the World Bank for an initial discussion.

He had hoped it would be a small meeting but somehow the World Bank came in force, led by a manager for the country. The World Bank country manager began the discussion:

> The manager started by pointing out that the World Bank focuses its support on ways to improve the lives of the poor and offer growth opportunities to the middle class. A key service that the World Bank provides to identify the best ways to do this is through technical analysis – for example, surveys of what people want, studies of issues and economic bottlenecks, brainstorming sessions, and so on.
>
> The World Bank prides itself on being a "knowledge bank," supporting a rich array of research based on global experience and local conditions, such as systematic country diagnostics, country economic memorandums, poverty assessments, private and financial sector assessments, and public expenditure reviews, in all of its client countries. This helps to zero in on the greatest needs of the poor and middle class. The World Bank also prepares a wide range of cross-country sector analyzes which can help to place a particular country's need within a global context. There is evidence that a positive correlation exists between effective operations and those sectors where World Bank studies have been done to identify needs and potential solutions (Hussain et al., 2018).

The finance minister's aide, having been earlier reprimanded for being too soft on the bank, decided it was time to correct this impression:

> Technical analysis can certainly help identify priority operations with the greatest potential to help the poor and build shared prosperity. However, the universe of analysis is far greater than that done at the World Bank, and the World Bank is often not good at using research prepared by others. It's important to look at a wide range of research, both internationally and nationally, before concluding on prioritization. Joint analysis prepared together by the government, the World Bank, and others would be particularly useful to build a shared vision of priorities.
>
> Another place the World Bank could do better is to learn from smaller pilot programmes provided by others and scale up the most effective pilots. Indeed, some of the smaller donors who do not have the finances to take an operation to scale have piloted successful (and unsuccessful) activities in the parliamentarian's constituency, and we need to learn from them.
>
> Finally, it is important not only to prioritize activities but to sequence them. For example, in many poor and fragile situations, early emphasis on overall basic financial management can be a prerequisite for sector-oriented

operations to be successful. A well identified operation should rely on a range of analysis and pilots, prepared both inside the World Bank and outside, to inform prioritization.

The finance minister expressed his appreciation for all of this technical analysis but decided it was time to redirect the conversation to ensure he would be in the driver's seat. Technical analysis is necessary but not sufficient for good operational identification. It cannot substitute for *country ownership*, which may be the single most important ingredient in development effectiveness (Partnership for Development: Proposed actions for the World Bank, World Bank discussion paper, 1998). Governments have their own sense of priorities, which are grounded in their own strategies and in a deeper understanding of how their societies function than external actors can grasp. Unless the World Bank and other lenders are open to listening and aligning their interventions with the countries they're trying to help, operations are unlikely to be successful.

The country manager wisely agreed that country ownership was essential and that the finance minister would be in the driver's seat. But he also noted that with ownership comes responsibility. It is equally important that the government in turn listens to, consults with, and takes into account the affected people, communities, and other stakeholders. The World Bank operates with requirements for consultations, and these are critical to ensuring that operations are demand driven. However, they cannot substitute for a government's own consultative efforts. Sometimes, governments that are less interested or able to hold their own wider consultations look to the World Bank to intermediate, but this is a mistake. Effective and sustainable operations are most likely to be identified when the government and the people being impacted are on the same page and feel they are being listened to by each other.

The finance minister felt this was a good time to consider the area in which the World Bank, compared to other institutions, was best placed to support an operation. It is important that the financing institution has a *comparative advantage* to finance the project – for example, expertise and experience in the area identified. In fact, the government might be tempted to ask the World Bank to take on a priority area that is best left to other international actors who have greater involvement in the area, either globally or at a country level. For example, supporting city police or cleaning up nuclear waste in some countries might be better left to specialized international agencies that have more expertise in these areas.

The World Bank country manager agreed that it is the overlap between technical analysis, government and local priorities, and the World Bank's comparative advantage that will identify the most promising operations:

Where there is no substantial overlap, effectiveness will be reduced, and both parties should consider walking away. However, in practice, focusing on the areas of substantial overlap or "sweet spots" is not that onerous – there are usually more than enough areas of substantial overlap for the financing available. This can be seen in Figure 2.1 below:

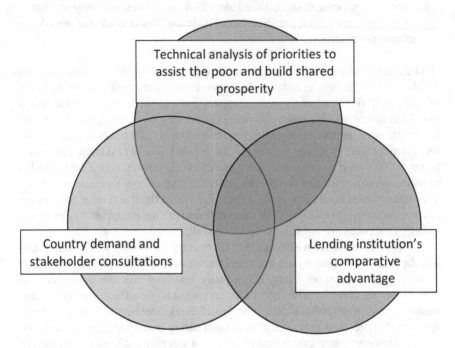

Figure 2.1 Identifying and prioritizing development operations

The country manager further noted that, while it is not a formal factor in identifying all priority operations, increasingly a fourth consideration is whether the activity contributes *to the global or regional public good*:

> Thus, for example, communicable diseases such as AIDS or TB or Ebola are of high importance to address from both a country and global point of view. Roads connecting countries can assist both internal and external trade. Most recently, the World Bank has increased its attention to assisting countries in managing refugee populations and to addressing climate change by reducing carbon emissions and supporting renewable energy.
>
> Sometimes, linking country needs to global priorities can even enable the World Bank to offer better terms (grants or more concessional lending) than might otherwise be the case. For example, in some global areas, donors have set up trust funds providing grant financing to encourage such efforts. However, not all operations have to have a global impact to be effective – for example, improvements in local street lighting and pavement can have a high payoff (such as that supported under the World Bank's FATA Urban Centers Project in Pakistan – World Bank, 2016).

After the meeting, the finance minister felt he had a better understanding of how to select the most promising areas for potential World Bank support.

The most effective projects would be those identified neither by the World Bank dictating to the country nor by the World Bank writing a blank check to the government. They would be identified based on the government's own plans to assist the poor and middle class, assisted by inclusive consultations with local stakeholders, and supported and refined by technical and practical analysis and pilots prepared by the World Bank and others. The government would also consider if the World Bank had the experience and expertise in the area identified, and if the activity contributed to both country and global/regional benefits.

Designing an effective operation

After identifying a priority area to be financed, the finance minister asked his aide to meet with the World Bank's experts to design the assistance in the most effective manner. At the meeting, the World Bank's programme leader was enthusiastic about the potential project and explained what her team could offer:

It is normal and appropriate for countries to look to the World Bank's *global knowledge and expertise* in designing operations. The World Bank's teams comprise both international experts as well as national experts who can help tailor these projects to the country context. For example, in designing irrigation projects, the World Bank has considerable experience it can bring to bear on how to ensure that both adequate drainage to reduce soil salinity and long-term protection of aquifers are built into the design.

The World Bank can also bring in experts from other areas to help focus the project. Broader studies prepared by the World Bank such as public expenditure reviews, poverty assessments, gender assessments, and private sector reviews can help shape the design of operations to be more effective and target those most in need.

At this point, the aide, having been properly warned by the finance minister to be wary of the World Bank taking over the design based on similar operations in other countries, decided to interject a note of caution:

The aide noted that government would be happy to listen to the World Bank's technical design team but would be careful not to rely on the World Bank too much in designing projects. It didn't want to become a bystander or peer reviewer for the project, but rather an active owner and designer of its own operation. It planned to get directly involved in project design, and not just at an overly parochial level (for example, the area to be covered by an agriculture project or ensuring that certain favoured institutions or consultants are used) but rather at a strategic level, and to consult with local communities and stakeholders for their input as well.

Perhaps other countries deferred to the World Bank on project design, given the power imbalance and need for financing. But this is a mistake. There is generally adequate *national talent and insight* that can guide project design within the government and local communities. An understanding of and integration with local customs and existing structures is invaluable for project effectiveness. Recognizing and accepting political necessities and what incentives work locally is essential.

The World Bank programme leader fully agreed on local ownership of design, and that this would not only help to create a better project but also ensure long-term sustainability. In thinking through the design together, she redirected the conversation to a good starting place. What results did the country want to achieve?

A key approach to designing effective projects is *starting from the results* that are desired. All too often, projects start with the physical thing desired – a road, or a school, or a health clinic. While these are important, starting from these and then considering what results can be achieved by them limits the potential design. If one starts with the result, say of increasing trade, this may lead to developing a road but just as likely the main constraint may turn out to be time-consuming border-crossing procedures or lack of supply chains.

Starting from the results can therefore avoid addressing an issue that is not the binding constraint. It is also important to pitch a result that is both achievable and a stretch, and then build a results chain (sometimes referred to as "a theory of change") to show how a specific activity can lead to this result being achieved.

The aide agreed that starting from results was important, though he knew that this was easier said than done since the parliamentarian would inevitably focus the conversation on tangible things. He wanted to make sure that while focusing on results, the project should be kept simple:

It is equally important from the start to keep *the KISS principle* (Keep it simple, stupid) in mind. International experts often love complicated operations with many components which allow them to show off their expertise. Because development, like life, is inherently complicated with many moving parts, it can seem more technically correct to design complicated operations than simple ones. There is also pressure from different global experts, each one with a favourite component.

However, the more complicated an operation is, the more problems that are likely to be encountered during implementation, particularly in countries where capacity can be limited. Simple is really better in the often-challenging world of development.

Both sides agreed to design a project that started from results and was not overly complex. The World Bank programme leader decided to turn to some other best practices in designing projects:

A key principle of effectiveness in design is to think *multisectorally* when preparing a project. Because both governments and international financial institutions are organized along sector lines (e.g., there is a Ministry of Education and a World Bank Global Practice for Education), it is very easy to design projects in silos. However, it is often the case that when design takes into account multiple sectors in a holistic manner, there is a richer, more effective result. In some areas, like trade or urban development or nutrition, it is generally accepted that a holistic approach is required to be effective. But even in the case of a more narrowly focused sector – for example, student learning – there are many factors that come into play, and it is important to understand the linkages before designing an effective project.

It is also useful to consider *connecting World Bank projects to global programmes*. In some areas, donors have set up global funds to target specific issues of global importance – for example, Education For All; the Global Fund for AIDS, TB, and Malaria; the Global Water Partnership, or the Green Climate Fund. These funds are referred to as "vertical" funds and bring attention and financing to critical areas but may have limited country-level capacity. World Bank operations can be particularly effective by linking its country-level capacity, knowledge, and involvement in institutional systems and reforms with these global funds to provide seamless donor support.

World Bank operations have been particularly successful when they support large *system-wide programmes* that are part of the country's core poverty alleviation strategy. Examples include (i) the National Solidarity Programme (NSP) in Afghanistan, providing villages across the country with support for services of their own choosing; (ii) the Benazir Income Support Programme in Pakistan, providing cash transfers based on objective poverty scorecards and through an internationally acclaimed "smart card" cash withdrawal system, to the poorest families in Pakistan; and (iii) the Bolsa Familia Programme in Brazil, which provides cash stipends to the poor through local governments, conditional on meeting certain requirements in education and health, and linking beneficiaries to other complementary services. In all three, the World Bank helped to strengthen systems and processes as well as financing to programmes fully owned by the government and designed to build social cohesion and reduce inequality on a countrywide basis (World Bank, 2017a, 2017b; IEG, 2014).

In many cases, including the above noted global connections or system-wide programmes, operations can be designed to have an important *catalytic* role – helping crowd-in other sources of funding, be they other international donors or private sector investors. For example, the World Bank's operation to support those affected by the 2010 floods in Pakistan – the Flood Emergency Cash Transfer Project – which supplemented US$125 million provided by the World Bank with US$355 million in co-financing provided by the United States, United Kingdom, and Italy (World Bank, 2014b).

Co-financing not only helps a country obtain more funding for a particular priority project than might otherwise be available from a single institution, it can help countries with weak implementation capacity to coordinate and pool financing from different sources using the same processes, rather than having to manage many smaller operations. An innovative "buy-down" form of co-financing was developed for Pakistan and Nigeria polio eradication projects whereby the Gates Foundation provided funds to cover the countries' World Bank loan repayments and interest obligations, essentially converting the loans into grants (Gay & Strickland, 2004).

The aide noted all of this information but repeated his concern that the project be kept as simple as possible, and that a multisector project might contradict this. In response, the World Bank programme leader chimed in:

> Projects that are simply a collection of components from different sectors certainly do contradict the KISS principle and should be avoided (these are often referred to as "Christmas Tree" operations). When an issue has a wide applicability across sectors, like gender, it is sometimes better to mainstream it into all projects. It is more challenging, but ultimately more rewarding, to design a simple intervention that impacts more than one sector. For example, supporting a simple conditional cash contribution to families that send young children to school can both help to strengthen social welfare and increase access to education by the poorest. A local roads project that is designed to be built with local labour provides communities with both jobs and accessibility and makes them more invested in the project.

The aide asked whether the World Bank takes into consideration a cost-benefit analysis in design. The programme leader was quick with a response:

> Bank policy requires investment projects to have a *cost-benefit analysis*, though the methodology and level of detail of each analysis varies from project to project. The cost-benefit analysis is usually based on net present value or internal rate of return under different scenarios or assumptions and compared to the counterfactual (e.g., the situation without the project). Where costs and benefits cannot be captured in financial terms – for example, environmental impacts – an economic rate of return is estimated.
>
> Cost-benefit analyses are more straightforward for infrastructure projects that have defined inputs and generate a clear revenue stream (for example, irrigation, highways, energy). Such analyses are more subjective in areas such as health or education, where results are long term and harder to quantify.

The aide appreciated that the World Bank prepared a cost-benefit analysis but was sceptical that this would help influence design:

> A review by the bank's Independent Evaluation Group concluded that "the Bank's use of cost-benefit analysis for decisions is limited because the

analysis is usually prepared after the decision to proceed with the project has been made" (IEG, World Bank, 2010). Because of this, the government would prepare its own cost-benefit analysis early in the process so it could help shape project design.

Finally, the aide asked whether World Bank–financed operations come with burdensome design requirements. The programme leader provided an explanation:

> World Bank–financed operations come with a number of *social and environmental safeguard and fiduciary requirements*, which inevitably add to design. For example, there are requirements that any operation involving land purchases (like hydropower dams or roads) be done in a consultative and open-market manner, with an easy-to-use complaint process. And any project impacting the environment must have a publicly available environmental assessment including mitigating measures. Increasingly, the bank also aims to make its operations gender sensitive (e.g., providing women with increased access to education, finance, healthcare, private sector opportunity, etc.).
>
> Countries may find these requirements onerous and look for less encumbered sources of financing, which is their right. At the same time, countries trying to improve their international image and standing may be happy to have these requirements, along with experts who will help them meet the requirements. An operation financed by the World Bank in a sensitive area can provide a very valuable "seal of approval."

Choosing the most appropriate type of instrument

Part of designing the operation is choosing the most appropriate type of instrument. The aide asked to be briefed on what options exist. The programme leader explained:

> Most operations financed by the World Bank are *traditional investment loans*, where a specific activity or set of inputs is financed. For example, a road or a hydropower dam or textbooks. Investment loans hold the benefit of being easy to understand. However, they can also get bogged down by the many steps needed for each "transaction" to take place, and when there are many small transactions the time and costs can get overwhelming. While all operations should be built up from results targets (as agreed earlier), it is sometimes challenging to translate the physical inputs and transactions into results, since there are many factors that go into achieving results outside of the project. For example, health clinics can be supported, but if the targeted poor are not incentivized to use them or if other parts of the supply chain like the availability of vaccines are not in place, the actual development result of improved health may not be achieved.
>
> For these reasons, a newer type of investment project – a *results-based operation* – was introduced about seven years ago and is increasing in popularity.

Under a results-based operation, funds are disbursed to the country or implementing agency based on their achieving specified results targets, rather than a regular investment operation that disburses against inputs. The specific activities to achieve these results are left up to the country. This has the advantage of providing maximum flexibility to the country in the type of activities and manner they are financed, while ensuring a tight focus on results.

The aide was intrigued by the idea of a results-based operation, not least because of its potentially simple design and it leaving the activities up to the government, but asked whether there was any analysis on the effectiveness of results-based operations that he could share with the minister. The programme leader responded:

Several studies have reviewed initial results-based operations and have found that most have been successful in achieving strong results (O'Brien & Kanbur, 2014; World Bank, 2015b). In Pakistan, a very effective results-based World Bank project was the Punjab Public Management Reform Program, which set targets for property registration, service institutions publishing results, and receiving citizen feedback. The project closed on time in December 2018, with all targets expected to be achieved or exceeded; this will be verified by a third-party monitor (World Bank, 2018a).

A more mixed assessment was had for the Punjab Education Sector II Project, which closed with a year delay in December 2016. The project was rated "moderately successful" with one set of key indicators, school achievement as per test scores, being achieved while another set of key indicators, net enrolment rates for primary school, actually decreasing both overall and for girls (IEG, World Bank, 2018a). A results-based operation is therefore no guarantee of full success and may require as much or more attention and effort.

It's therefore important to recognize both the potential of results-based operations and the challenges. Perhaps the biggest challenge is ensuring that a results target is set at the right level. Setting it very high can provide stretch incentives but may also be unfair since high-end results can depend on many factors, both internal and external to the country. In the other direction, it is important to resist incentives to set the bar too low on the results chain, which will be easy to meet and obtain disbursements on schedule but dilutes the development benefits.

A second important challenge is to identify results indicators that are both easily measurable and transparently verifiable. It does no good to pick a results indicator that is not regularly monitored. It is also helpful to have an independent and/or transparent process to reduce potential incentives to manipulate results.

A third challenge is that the means to achieving a result can be as important as the ends. Since results-based operations leave specific procurement up to the implementing agency, corrupt or heavy-handed practices may occur with little of the World Bank oversight that would otherwise exist under regular investment operations. A fourth challenge is establishing an appropriate amount to be disbursed against a particular result.

In sum, results-based operations can be effective, but only if results targets are identified that are easily measurable and high enough on the results chain and at a level to matter. It is equally important for the implementing agency to have well vetted and professional procurement and management practices in place to achieve results.

The programme leader suggested that both regular and results-based operations can benefit under the right circumstances from *technical assistance* being provided to the implementing agency either as part of the operation or in parallel. Since many implementing agencies have limited capacity, a small amount of technical assistance is money well spent to help implementation go smoothly. As predicted, the aide recognized the usefulness of technical assistance but expressed some wariness:

Some technical assistance can certainly help implementing agencies to improve pre-identified skills or knowledge gaps. However, it can also be overused. For example, study missions can end up as junkets without tangible benefits. In addition, in countries where rotation among civil servants is high, strengthening staff capacity only to have them leave after one or two years can limit its usefulness (though they may use these skills in another agency).

There is also a tendency to hope that technical fixes can resolve policy challenges, which is rarely the case. The Tax Administration Reform Project (TARP) from 2005 to 2012 in Pakistan sought to use technical assistance to strengthen tax administration and improve automation. However, the project did not increase tax revenue to GDP and the Implementation Completion Report (ICR) recognized that "tax policy and other legal changes need to be taken in parallel with institutional capacity development to achieve full benefits from planned reforms" (World Bank, 2012).

The programme leader explained two other products that the World Bank offers: Countries have the option to obtain financing in the form of policy-based operations. These comprised roughly one-quarter of World Bank financing over the past decade (World Bank, 2015a). In these operations, funds are disbursed based on the country approving agreed pre-specified policy reforms, often requiring administrative approval and/or Parliamentary submission. Where countries are contemplating policy reforms, these operations can provide an extra incentive and are particularly useful in situations where budget deficits or weak reserves have increased the need for cash infusion at a macro level. More recently, the World Bank has begun to offer policy financing on a "contingent" basis (known as a deferred drawdown option) to address unforeseen natural disaster shocks. Under this option, a country would establish policies and plans for a disaster, but the funds would only disburse if there was such a shock (and until then, the country would pay only a small commitment fee).

The biggest effectiveness challenge for policy-based operations is assessing the degree to which countries are really committed to the agreed upon policy reforms. If countries only agree to policy changes out of a need for funds, this can lead to ineffective and reversible policy changes which follow the letter of the law but not the spirit. For example, a government not committed to real trade reform may agree to reduce formal tariffs but at the same time place non-tariff restrictions to protect certain goods. As these operations are normally designed as a single payment which disburses within a year, sustainability can also be an issue. While the bulk of the supported reforms are normally sustained for at least several years, it is also true that the politics of the country may change, and a new government may reverse the reforms with little consequences.

A final alternative is the use of *guarantees*, which may be used for either investment or policy operations. These can be helpful where, for example, a country is seeking to attract external investments, and private investors exist but need assurance of a purchase price to be profitable, or assurance of repayment by the government. The World Bank, like other international financial institutions, has the ability to offer guarantees, but its use has been limited since investors may either not be available or may not want to make their incentive inflection points known. Further, countries sometimes prefer direct loans to guarantees since the use of guarantees partially counts against the World Bank's lending limits to a particular country.[8]

Setting the terms of the operation

The aide turned to the potential terms as he felt this would be an important element in thinking through the project. The programme leader provided a general brief:

> The basic financing terms are determined according to the relative wealth of each country – all grants, a mix of grants and long-term concessional loans (roughly 2% currently), long-term concessional loans only, a mix of long-term concessional loans and market-based loans (roughly 3% currently), or long-term market-based loans only. But these rough ranges can hide a further range of options – including choice of currency, fixed or floating, and differences in maturities or grace periods – which also vary according to the relative wealth of the country.
>
> Normally, after the project is designed but before it is approved, the World Bank provides a technical briefing and seminar on the different options the country has to choose from. Choosing the right terms can help to match a country's debt profile with its revenue stream profile so as to reduce risk. However, World Bank staff are prohibited from making recommendations lest they reduce a country's ownership of decisions and/or be blamed for any adverse outcomes.

The aide worried that it would be time-consuming to review these options, as few government officials want to be held accountable for adverse outcomes and adding to the country's debt. At the same time, such a high level of scrutiny on terms was unlikely to add much value and might actually take attention and time away from the selection and design of the project, where government involvement could add more value.

The fact is, whatever option is chosen, international financial institutions like the World Bank are able to provide funds at a very competitive rate, normally close to or below those associated with AAA credit ratings. Just as importantly, there is no objectively "cheaper" alternative – all options for a particular country are designed to have a neutral overall impact. And the specific operation would be just one of many loans in a country's debt profile, and therefore would be unlikely to make a systemic difference in the risk.

Staffing the operation

The minister of finance knew that the quality of staffing could make or break a project, and that the time to think about staffing was well before the project launch. A poorly designed project can be made to work by experienced staff, while many well-designed projects languish if not staffed well. The minister also felt it would be a mistake to rely too heavily on World Bank staff for project implementation. Therefore, he asked for the head of the identified implementing agency to brief him on their staffing plans. The agency head met him in his office and indicated the following:

> The implementing agency confirmed that they were in the process of hiring a competent team – including experts in operations management, procurement, financial management, Monitoring and Evaluation (M&E), and technical officers skilled in the issues being addressed – all before launching the project. The agency also recognized that there would need to be outreach to the effected communities, stakeholders, and citizens – for example, through an advisory council.
>
> However, the agency head needed the minister's agreement to use government budget funds to hire these staff members before the project was launched. He noted that if the government put its own funds in for this purpose, it could be reimbursed from project funds once they were available. Countries also have the option of taking an advance of World Bank funds through a Project Preparation Facility (PPF), though in practice the potentially heavy internal approval process for this within the country (often similar to the operation itself) can make it less attractive. He cautioned that without having access to either government budget funds or a PPF at this stage, the project start could be delayed by more than a year.

The minister agreed that he would permit budget funds to be used for this purpose. At the same time, he asked the agency head to be very careful to select staff and plan outreach in a transparent manner to reduce the likelihood of them being chosen based on political or family connections, and thus not having the needed skills. He asked his aide to be kept in the loop on the hiring and outreach process, also aware that World Bank staff would be checking to make sure a professional process of hiring was used. He further asked the agency head to make sure the team was kept largely intact once selected. The agency head agreed:

> Once a professional team has been selected, avoiding excessive turnover becomes essential. Some civil services provide for staff to rotate every year or two, which means that projects can be constantly delayed or disrupted when those with more experience leave and new staff members have to be hired, which in some cases can take six months or more. Ironically, international financial institutions can themselves be the cause of excessive rotation by encouraging implementing agencies to hire experienced staff from another ongoing project (possibly from a project financed by a different financial institution or even a different sector within the same institution).
>
> It would be equally important to keep the team intact without ring fencing it from the rest of the implementing agency – for example, avoiding the creation of a legally separate Project Implementation Unit. In the near term this can enable the project to be insulated from political concerns and to offer higher pay and therefore attract higher quality staff. But in the longer term, it detracts significantly from both country ownership and integration with the rest of the activities of the implementing agency and communities. It can also create negative reactions and incentives for regular agency staff who are paid less. This does not mean there would not be dedicated staff for the project. There would be. But they would be integrated into the implementing agency as much as possible.

Finally, the minister asked how the implementing agency would interact with World Bank staff assigned to the project, to make good use of their expertise without ceding control. The agency head responded:

> All project activity would be done by the implementing agency. It can be tempting to lean on World Bank staff to do some of the work (for example, to write a procurement request for a proposal). World Bank staff eager to show results can also be tempted to offer such work. However, while this can make things work faster in the short term, it is a long-term mistake. It is much better for the long-term success of the project to allow implementing agency staff to remain on the hook for all aspects of direct project implementation to build ownership and knowledge.
>
> Nevertheless, World Bank supervision staff can play an important support role, bringing a great deal of external review and expertise to bear, as well as training and problem solving when there are issues. There is a strong correlation between project effectiveness and World Bank staff who are experienced and who stay with the project from design to completion (Hussain et al., 2018).
>
> World Bank staff are most effective when they comprise a team of experts in different fields, as well as a good mix between national staff, international

staff, and consultants (to ensure access to both global knowledge as well as local conditions). Senior World Bank officials may also be called on to help if there are issues that cannot be resolved at the working level.

The conversation reassured the minister. He would need to monitor the implementing agency regularly to ensure the project was professionally staffed, but he was glad to find that project control would be kept within the implementing agency, with World Bank staff being used for expert support and review.

Ensuring timely flow of funds during implementation

It is not uncommon, and it is extremely frustrating, to have a well-designed and well-staffed project where the funds are simply not being disbursed. Before the project launch, the implementing agency project team and World Bank supervision task team met to discuss how they would ensure the timely flow of funds:

Unfortunately, there are too many projects that stay on the books for much longer than anticipated when designed. Not only does this delay development impact, but as situations and costs shift over time the net benefit may also decline. Where projects do not start to disburse significant funds within six months to a year of being approved, this signals an urgent need for review. If funds stay blocked for a long time even with the team's and senior official's attention, cancellation should be considered. Perhaps the project was actually not as well-designed or country-owned or as well-staffed as was thought. But there are also often unforeseen blockages that can and should be addressed.

One blockage that can occur is when investment operations require the government to provide additional *funds from its budget* (known as "counterpart" funds) alongside those available from the World Bank. This can be helpful both to ensure ownership and to cover activities such as staffing costs beyond the core team or inputs other than core project inputs. Nevertheless, because such funds come from the government budget they need to be appropriated by Parliaments and allocated to implementing agencies. And government budgeting can all too frequently be held up for totally unrelated technical or political reasons. When counterpart funds are held up, follow-up from both World Bank and government staff is needed; if possible, it's also helpful to identify a champion in the budget department who can assist.

A second key area of blockage can occur during the *procurement process*. The World Bank requires transparency and fairness in the procurement process in investment operations it finances. Procurement can be quite involved and requires experienced, well-trained project staff who know how to navigate the process – for example, call for proposals, shortlisting, selection criteria, selection committee, and so on. In addition, the implementing agency project manager may feel political pressure or financial incentives to award contracts in a biased manner. Complaints and related court cases have been known to hold up procurement for years. Sometimes even just the potential for complaint can hold up the process by a risk-averse manager. The key is really to make sure there are well-trained professionals who understand procurement processes from the start, since once the procurement train is derailed, it is often very difficult and time-consuming to get it back on track.

Compliance with *environmental and social safeguards* can hold up projects if not well-designed from the outset. Even if they are well-designed, actions like land purchases often take longer than anticipated to meet safeguard requirements, but with careful attention these issues can be successfully addressed during implementation. For example, in Bulgaria a transportation project was delayed due to slow land acquisition and chance archaeological finds; but although it was completed two years later than anticipated, it was rated as having a satisfactory outcome (World Bank, 2014a). Close interaction between World Bank supervision staff and implementing agency project staff to understand and anticipate these challenges is essential.

Implementing agencies may also have *capacity constraints*, which can lead to weak disbursements. "Capacity constraints" is a very general term that can refer to inadequate staffing, poorly trained staff, unclear or ineffective processes, or simply a lack of initiative or will. Bank support, training, and/or technical assistance can be helpful but this must be specifically targeted and should support but not substitute for agency staff.

The World Bank team further indicated their own supervision processes:

All World Bank projects are supposed to be *supervised* every six months, and a public supervision report of any issues or blockages posted. In fact, this requirement is an echo from decades ago when projects were supervised largely by Washington-based staff who would travel on mission to supervise a project. Today, most projects have task team leaders, co-leaders, or key members who are field-based and ought to be continuously supervising projects. The answers to solving implementation issues rest with the project teams in the implementing agencies and not in pushing paper in the World Bank office.

The more World Bank teams interact with implementing agencies and clients (and are empowered to work with implementing agencies to make decisions) the more likely issues will be addressed, and the project will be effective. Sometimes, particularly in conflict-affected areas, World Bank teams may be limited in their ability to visit sites directly due to safety reasons. In these cases, the World Bank sometimes employs satellite imagery or local third-party monitors who can more easily gain access to development sites and provide up-to-date data.

Although senior World Bank and government officials can be brought into the picture at any time, in most cases a regular *portfolio review* is held annually or more frequently if needed. These are normally co-chaired by senior Word Bank staff and a high-ranking government official. During such a portfolio review, individual project teams from both the implementing agency and the World Bank raise policy or fund flow issues that need high-level attention to resolve, as well as more generic issues (such as staffing and procurement problems). In extreme cases, disbursements can be suspended, or funds cancelled if issues remain unresolved. Sometimes, just the threat of suspension or cancellation is enough to motivate the necessary action.

The implementing agency and the World Bank teams agreed that supervision and joint portfolio reviews can only resolve blockages if the two teams are candid about issues and ways to solve them. Sometimes teams put off raising difficult issues so not to bother their superiors or to avoid difficult project restructurings, but the fact of the matter is that the longer an issue or blockage is not addressed, the harder it can be to resolve.

Monitoring, feedback, and reporting

How do we know if a project is successful or not? Clearly if funds are not disbursed, it cannot be successful. But even if project funds are disbursed, success is not assured and must be measured. World Bank staff met with their counterparts and explained the process as follows:

The key to effective monitoring comes in establishing good *results targets* from the start, during the design phase. Effective results targets are those that are meaningful, measurable (with good baselines from the beginning), achievable (but not too easily), and attributable to the project in question. For example, in education, it is fairly easy to measure school attendance, but this may provide a false positive which does not translate into improved learning (due to a high level of teacher absenteeism, for instance). A false positive could also emerge from setting a result target impacted by uncontrollable externalities. For example, setting a result target of less expensive electricity for an energy project sounds good. But this could be a misguided goal if the price of energy is more determined by international geopolitics than national improvements.

The implementing agency is primarily responsible for collecting data against results, and regularly reporting them to both senior government officials as well as World Bank supervision staff. All too often results monitoring takes a back seat to the activities being financed, but this will likely reduce the ability to take corrective action early on.

At least every six months, the World Bank staff file a *publicly available supervision report* that indicates progress against agreed results and can identify implementation issues that need to be addressed. The best practice is for bank staff to also provide implementing agencies with a longer report of their mission, called an aide-memoire, particularly for more complex or challenging projects. However, aide-memoires can get overly long and technical, and are too often watered down through internal clearances – the best aide-memoires are frank assessments provided directly between Bank project teams and their implementing agency counterparts, starting from progress against results and providing a quick list of three to five issues to follow-up on with timetables and responsibilities.

At the *midterm*, the World Bank would normally undertake a longer mission that would assess whether any project restructuring is needed, including changes in results targets, activity components, and closing dates. Changing results targets can be justified as projects do change in approach, and new data sources can emerge. Adjustments to results targets can be a positive development as sometimes operations do need to adjust to changing circumstances.

However, it is equally possible for projects to fall short of their initial expectations and rather than closing them down or doing radical surgery, the results targets can be declared to be unrealistic and made easier to achieve. In evaluating project effectiveness, it is therefore important to compare actual results to both the original and revised target, and to make a judgement whether any change in the target is justified.

In addition to regular supervision and midterm reviews, the World Bank requires that all operations it finances be subject to *annual audits*. It is important to ensure that the auditing firm selected is objective and professional. Sometimes government audits should be supplemented by private audits. For the most part, audit reports simply certify that funds were spent as they were intended. However, sometimes audit reports can identify significant issues either with the procurement processes, fund utilization, budget management, or other accounting processes that require corrective action.

It is good practice for projects to include *citizen feedback* during implementation in order to measure responsiveness to clients. A common example would be a simple client survey measuring satisfaction with the service provided. A fairly common mistake is for the survey questions or audience to shift between the initial survey and the survey at the end of project, thus leading to inconclusive or fairly useless information. A better approach would be to develop a system that provides regular feedback (say, every six months) to enable reliable tracking and time for project adjustments as suggested.

All projects financed by the World Bank have a *completion report*, which assesses effectiveness and sustainability. In addition, the World Bank's Independent Evaluation Group (IEG) reviews completion reports to provide a separate evaluation of how well the project met its agreed objectives and results (efficacy), as well as project relevance and efficiency. Where there are differences between the team's completion report and IEG's assessment, and between ratings in completion reports and ratings provided in ongoing operations, these are referred to as a "candour" gap and suggest that the country teams may need to be tougher in their assessments. The most important element of these reports is the "learning" aspect to influence new projects. All completion reports include a section on lessons learned, and these should be systematically used to influence the design of new projects.

Conclusion

The finance minister and his team compared notes and reflected on what they had learned about working with the World Bank, and the questions that needed to be answered in order to determine whether a World Bank "package" would be effective.

They knew to ask whether the World Bank had performed an analysis of and had experience in the areas in question to help prioritize and sequence activities, but to also be careful to take other analyses into account and not to largely cede project selection and design to the World Bank. They knew to ask if local, political, and cultural sensitivities, which are critical to the success of an operation, had been taken into consideration. They knew to make sure that the operation was designed with a focus on results not just inputs, and not to be overly complicated.

They knew that World Bank operations could come with challenging social, environmental, and fiduciary requirements, but that this can also give sensitive operations a stamp of approval. They knew to check if the operation could be designed so that they could bring in other sources of funding, including grants. While World Bank operations come with professional advice and staffing, they knew that these cannot and should not substitute for building up the government's own operational institutions and staff. They knew that World Bank operations could help to bring transparency to results monitoring, and the attention of high-level government officials could help to troubleshoot solutions when funds are held up for bureaucratic reasons, but also that this required honest and early discussions between the teams to be effective.

They concluded that the single most important thread throughout any "package" is that the national government and the local community need to take ownership and management of the project – with World Bank efforts providing support but not replacing those of the country. This is true in the identification and design of projects, in staffing, in monitoring results, in budgeting, and in citizen consultations. Because of the richness of the "package," it is all too easy for a country to let the World Bank take leadership. But ultimately this is a mistake that will play itself out in countless implementation difficulties and most likely an ineffective and unsustainable operation.

They also concluded that the second most vital thread throughout any "package" is the importance of focusing on results. This is true of project prioritization and design, but also important in implementation and supervision. Identifying a measurable, achievable but stretch results target, using it to identify and design the project, and monitoring the results target laser-like throughout the process (being careful not to change it without need) provides both the focus and the incentives for project effectiveness.

At each stage, the different parts of the World Bank "package" need to be recognized and judgement used to ensure that the project is well-chosen and well-developed – from identification and prioritization, to project design, to lending choices (investment, policy-based, results-based, guarantees, technical assistance), to staffing, procurement, auditing and monitoring, and completion. If most or all of the factors contributing to effectiveness align, the government may indeed be justified in obtaining World Bank support for an operation. In these circumstances, the returns would likely far outstrip the cost of the loan. If most or all of the factors do not align, World Bank financed operations would be unlikely to succeed, and the country would be taking on debt for activities that would probably not pay off.

Notes

1 This discussion only looks at the arms of the World Bank that provide finance to governments – these are the International Bank for Reconstruction and Development (IBRD) for higher-income developing countries and the International Development Association (IDA) for lower-income developing countries. The other branches of the World Bank Group – International Finance Corporation (IFC), Multilateral Investment Guarantee Agency (MIGA), and International Center for the Settlement of Investment Disputes (ICSID) – provide support to the private sector and/or have a different dynamic.
2 The author is a retired World Bank staff member with 35 years of experience working at the World Bank in operations and strategy. The views expressed herein are his own and do not necessarily represent the views of the World Bank and its affiliated organizations, or those of the executive directors of the World Bank or the governments they represent.
3 Extreme poverty is defined as those living on under US$1.90 per person per day, and the goal is to reduce this from the current level of 10 percent to less than 3 percent by 2030. Shared prosperity is defined as the growth in income of the bottom 40 percent of each country.
4 Organisation for Economic Co-operation and Development (OECD) data indicates 20 developing countries where official development assistance exceeded 10% of their Gross National Income (GNI) in 2016, and another 20 where it amounted to between 5% and 10%.
5 Amounts are only financing to governments – in other words, excluding flows from private sector institutional arms.
6 Non-concessional loans are the amounts committed under the IBRD window, and concessional loans and grants are committed under the IDA window as per the World Bank's Annual Report. In practice, the amounts indicated undercount the total financing provided by the World Bank and the share of grants in it, as in addition to its own funds the World Bank administers a large number of country-specific and sector-specific trust funds provided as grants on behalf of donors.
7 All rules have exceptions, and where there is civil war and a limited functioning government, the bank can provide funds directly to NGOs.
8 For the bank's concessional operations, guarantees count four to one against country allocations. For non-concessional operations, the bank counts guarantees one-to-one against its capital, but at a country level can count four-to-one. Leverage can also be increased through the degree to which guarantees only cover part of the risk.

References

ADB (Asian Development Bank). (2018) *Annual Report*. Manila, ADB.

AfDB (African Development Bank). (2018) *Annual Report*. Abidjan, AfDB.

Gay, A., & Strickland, S. (2004) Buy downs' helping to eradicate polio, *Alliance Magazine*, September.

Hussain, M. Z., Kenyon, T., & Friedman, J. (2018) *A New Look at Factors Driving Investment Project Performance*. World Bank DEC Policy Research Talk, September, Washington, DC, World Bank.

IEG (Independent Evaluation Group, World Bank). (2010) *Cost Benefit Analysis in World Bank Projects*. Washington, DC, World Bank.

IEG (Independent Evaluation Group, World Bank). (2014) *Brazil Country Program Evaluation FY2004–11*. Washington, DC, World Bank.

IEG (Independent Evaluation Group, World Bank). (2018a) *Punjab Education Sector II Implementation Completion Report (ICR)*. Washington, DC, World Bank.

IEG (Independent Evaluation Group, World Bank). (2018b) *Results and Performance of the World Bank Group 2017*. Washington, DC, World Bank.

Kamel, N., Cholst, A., Funna, J. S. A., Guerrero, P., Hilton, R., Mitchell, J., & Nishio, A. (1998) *Partnership for Development: Proposed Actions for the World Bank – a Discussion Paper*. Washington, DC, World Bank.

O'Brien, T., & Kanbur, R. (2014) The operational dimensions of results-based financing, *Public Administration and Development*, 34(5), pp. 345–358.

Winters, M. (2010) *Targeted Aid and Capture in World Bank Projects*. The Annual Meeting of the American Political Science Association. Washington, DC.

World Bank. (2012) *Implementation and Completion Results Report on the Pakistan Tax Administration Reform Project*. Washington, DC, World Bank.

World Bank. (2014a) *Implementation and Completion Results Report for a Second Trade and Transport Facilitation Project for the Republic of Bulgaria*. Washington, DC, World Bank.

World Bank. (2014b) *Implementation and Completion Results Report for the Flood Emergency Cash Transfer Project for the Islamic Republic of Pakistan*. Washington, DC, World Bank.

World Bank. (2015a) *Development Policy Financing Retrospective: Result and Sustainability*. Washington, DC, World Bank.

World Bank. (2015b) *Program for Results: Two-Year Review*. Washington, DC, World Bank.

World Bank. (2016) *Implementation and Completion Results Report for the FATA Urban Centers Project for the Islamic Republic of Pakistan*. Washington, DC, World Bank.

World Bank. (2017a) *Implementation and Completion Results Report for the National Solidarity Program III for the Islamic Republic of Afghanistan*. Washington, DC, World Bank.

World Bank. (2017b) *Implementation and Completion Results Report for a Social Safety Net Project for the Islamic Republic of Pakistan*. Washington, DC, World Bank.

World Bank. (2018a) *Punjab Public Management Reform Project Implementation Status and Results Report*. Washington, DC, World Bank.

World Bank. (2018b) *Annual Report*. Washington, DC, World Bank.

3 SDG tracking through SSC for better SDG spending

Hasanuzzaman Zaman
and Syed Sajjadur Rahman

South-South Cooperation (SSC) focused on knowledge sharing for tracking SDGs implementation, can help improve SDG spending. This chapter will be the first to document how the Global South can share knowledge on SDG through development cooperation. The policy transfer framework will anchor the discussion on the SSC which has not yet been done to date at the academic level. Bangladesh has an SDG tracker, much like the UN's global SDG tracker, which is used by the Finance Division and Planning Commission for budgetary decisions. A discussion on how the SDG tracker is being used by the government of Bangladesh will set the scope for analyzing policy transfer between a least developed country (LDC, i.e., Bangladesh) and a middle-income country (Peru). Bangladesh's design and use of the SDG tracker and the policy transfer with Peru can illuminate new ways for improving SDG spending through existing administrative arrangements.

Keywords

Policy transfer, new public governance (NPG), collaboration theory, communicative rationality, South-South Cooperation (SSC), development cooperation, aid effectiveness, SDG tracker

> *The model of development cooperation of the 1960s and 1970s which saw a world of the North transferring technology and models of development to the South is long gone. Much of the innovations that drive development today have their origins in the South.*
>
> – Achim Steiner, UNDP administrator and secretary-general of the Second High-Level UN Conference on South-South Cooperation, known as BAPA+40

Introduction and objectives

Countries in the so-called Global South are benefitting from rising South-South Cooperation (SSC) facilitated through trade, investment, knowledge, and technology transfer (UNDP, 2016). For example, there have been positive growth spillovers from China to its trading partners, and there has been the supply of affordable medicines and information and communication technology (ICT) products and services to countries in Africa from India (UNDP, 2013). The increase in the number of South-based donors has changed the traditional North-South aid landscape (Rahman & Baranyi, 2018).

The concept of SSC first emerged in the Bandung Conference back in 1955. The Buenos Aires Plan of Action (BAPA) in 1978 on technical cooperation gave SSC an international framework. Further institutionalization of the concept took place during the Millennium Summit in 2000 and the Nairobi Conference in 2009. The Busan Declaration in 2011 prioritized SSC in development cooperation. The Second High-level United Nations (UN) Conference on SSC was held in Buenos Aires, Argentina, in March 2019. It noted how development cooperation architecture is undergoing changes as a result of the "rise of the South" (UNDP, 2013), and concluded with a declaration to renew the global commitment to leverage SSC as a means to contribute to the achievement of the 2030 Agenda for Sustainable Development Goals (SDGs).

During the Second UN World Data Forum in Dubai in October 2018, the governments of Bangladesh and Peru signed a Memorandum of Understanding (MOU) to share knowledge on the design and use of their national-level SDG trackers (Askari & Luna, 2019; UNOSSC & SSN4PSI, 2019). This MOU foreshadowed the conclusions of the Buenos Aires conference. The UN Office for South-South Cooperation's (UNOSSC) compendium had featured the SDG tracker of the Bangladesh government as an illustration of a good measurement mechanism (UNOSSC, 2018). Bangladesh and Peru are now discussing ways for operationalizing their MOU.

The Bangladesh-Peru MOU concerns an area of increasing global interest which is about the measurement of SDGs progress. This measurement and evaluation process is complex given the 17 goals, 169 indicators, and 230 targets that need to be monitored. It is especially difficult for Southern countries because of a lack of financial and human resources. The measurement is also made more difficult given the development status differences between the countries and the impossibility of creating a homogeneous framework. The Bangladesh-Peru MOU represents an attempt by two Southern countries to learn from each other in order to design SDG tracking and measuring instruments. The ambition is to become more effective in their development efforts and also to economize on the costs of regularly updating and improving the instruments themselves.

The hypothesis of this chapter is that SSC, focused on the SDG tracking knowledge transfer, can help improve effectiveness of SDG spending. The analysis and recommendations here are directly relevant for developing countries aiming to improve budgetary effectiveness by using instruments like SDG trackers to target resources for accelerating progress towards their SDG targets. This chapter analyzes the ways SSC can be formalized as a policy transfer "learning" mechanism for improving decision-making processes in aid and public expenditure. The interest here is more on the *process* aspects of how policy transfer can take place under SSC using the Bangladesh-Peru MOU as an example. We cannot confirm whether the MOU has led to improved SDG spending yet because it is in the early stages of implementation. Nevertheless, this is the first analysis of an SSC type of policy transfer combining conventional public policy ideas and frameworks. Besides the new academic contribution, this chapter offers lessons on how such policy transfer mechanisms can be established and leveraged in the Global South for better SDG spending.

Methodology

The analysis in this chapter uses a mix of primary and secondary sources of data and information. High-level officials from the governments of Bangladesh and Peru with direct knowledge and experience of the MOU's design, process, and outcomes were interviewed, along with two representatives from the Children's Investment Fund Foundation (CIFF).[1] They provided the authors access to the signed MOU, along with the latest implementation roadmap being discussed by both countries, and offered valuable insights which strengthened the analysis. CIFF played a critical role in not only connecting the two governments, which are nearly 18,000 kilometres away from each other, but also in organizing regular bilateral meetings that led to the finalization of an MOU, and ultimately its signing by Bangladesh and Peru. We reviewed the policy transfer literature and new public governance (NPG) in order to analyze how a collaborative form of rationality can guide a policy transfer learning process. Documents from the UNOSSC and think tanks as well as government reports of Bangladesh and international agencies were reviewed in order to examine the rationale and utility of SSC in promoting a "learning" type of policy transfer mechanism for SDG tracking.

The analysis combines policy transfer processes with NPG's collaboration theory. This has not been previously done, even though policy transfer fits well with NPG's pluralist framework. We extend this analysis within the context of the SSC type of policy transfer mechanisms and challenge the presumption that policy transfer is primarily a North-South phenomenon. Using the Bangladesh-Peru example, we illustrate a policy transfer process where collaboration is about *communicative thinking* leading to collaborative rationality to guide decisions (Healey, 1992, 2010). The similarity in needs and abilities is identified by actors, and the emphasis is on the role of "visionary leaders" in such discussion processes including high-level and mid-level bureaucrats, politicians, and non-government experts (Crosby et al., 2010).

Research questions and chapter layout

We investigate four broad questions:

- What is the attraction of Bangladesh's SDG tracker in terms of increasing SSC utility as a policy transfer learning mechanism?
- How can policy transfer and NPG theories be combined to explain the type of policy learning taking place in the South?

- What conditions make South-South policy transfer possible and in what form do such collaborations take place?
- How can external organizations (like international aid agencies and philanthropies such as CIFF) help to promote SSC for the improved effectiveness of SDG spending?

In the second section, we present a brief discussion on Bangladesh's SDG tracker to underline the potential of such ICT instruments in improving SDG-related expenditures. This discussion provides the rationale and utility for leveraging SSC. The third section combines policy transfer with NPG's collaboration theory in explaining an SSC type of "learning" policy transfer mechanism. The fourth section analyzes the process and contents as well as current status of the signed MOU between Bangladesh and Peru based on online interviews with Peruvian officials and CIFF's representatives, and in-person interviews with officials from Bangladesh. The fifth section concludes with some key messages for leveraging SSC as a way for improving effectiveness within the SDG context.

Bangladesh SDG tracker: a "development mirror"

A number of countries have developed monitoring and reporting instruments to report on their SDG statistics. For example, the United States recently updated its SDG platform launched in 2017 (Wagner, 2018). The US site was built as an open-source tool and the US government is collaborating with the United Kingdom and Ghana on creating national reporting platforms (ibid.). The government of Canada has created a new SDG Data Hub as a one-stop online data repository to track the country's progress.

The SDG tracker of Bangladesh: rationale and utility

Bangladesh's SDG tracker is also known as the "development mirror" of the country. UNOSSC (2018) commended Bangladesh's achievement in implementing MDGs and noted that the country was poised to do better in achieving the SDGs as well. However, unavailability of timely and quality data had, in some cases, hindered the attainment of MDGs (ibid.). The SDG tracker is being used as a timelier and more effective tool to track the implementation of SDGs. In the pre–SDG tracker period, data produced by various agencies were maintained in different silos and needed to be assembled manually for decision-making purposes. The Bangladesh Bureau of Statistics (BBS) under the Planning Commission produces most data on aggregate development indicators. The data is mainly collected from the records of ministries and their divisions at the national level, and from directorates and other government agencies at the field level.

The Access to Information (a2i),[2] an innovation centre of the government linked to the Prime Minister's Office (PMO) until 2018, in collaboration with the General Economics Division (GED) of the Planning Commission and the BBS, designed the SDG tracker of Bangladesh (a2i, 2018). The objectives were to do the following:

- Create a data repository for monitoring implementation of SDGs and national development goals set under the five-year plans.
- Facilitate tracking of progress against each goal and target through multiple visualization schemes.
- Improve situation analysis and performance monitoring.
- Create an environment of healthy competition among various organizations to achieve SDGs.
- Enable predictive analysis for achieving the goals within the set time frame.

The cost of developing Bangladesh's SDG tracker was US$62,000, and an additional US$50,000 was allocated for its maintenance during the 2018–2020 period (UNOSSC, 2018). According to our interviews, after 2020, the maintenance contract will be extended and maintained by a2i.

The Bangladesh SDG tracker will be measuring 40 indicators across the 17 SDGs, which have been prioritized by the PMO. The tracker can be customized to any language and is easily transferrable. It has two components: the SDG portal and the SDG dashboard (please see https://a2i.gov.bd/sdg-tracker/; a2i, 2018). The former is a crucial part of the SDG tracker; it enables policymakers, government agencies, private sector entities, civil society organizations (CSOs), international organizations, academics, researchers, and above all, citizens to track year-on-year progress against each indicator and target. Users can create their own visualization depending on their interests. They can also retrieve data related to MDGs as these data have been archived in the SDG portal. The SDG portal can be used to track the progress of the ongoing national devolvement plan.

The purpose of SDG dashboards for individual ministries/divisions and agencies is to consolidate available data for each SDG and compare it visually against performance thresholds. For example, the tracker can show progress on a number of the SDGs in specific target-based terms. The resulting dashboards highlight areas where a ministry needs to make the most effort and allow government agencies to prioritize early action, understand key implementation challenges, and identify gaps that must be filled in order to achieve the SDGs. The dashboard can also provide a visual representation of the whole SDG tracker system showing the flow chart of the progression of SDG indicators from definitions to the final reporting process.

At the administrative level, a high-level principal coordinator position was created in 2018 at the PMO to head up a committee on SDG affairs and for overseeing the SDG tracker's use by the government. The GED of the planning commission serves as the secretariat and also the focal point. One responsibility of the principal coordinator is to popularize the SDGs in order to create awareness among citizens about the SDG tracker. An SDG localization effort is also underway. The Planning Commission has published a Bengali version of a report on SDG targets and indicators, which was distributed across all public offices all over the country. The GED has formulated a mapping of the ministries/divisions to identify responsible agencies for achieving the SDGs (GED, 2017). Bangladesh has instituted new policies, processes, and practices to integrate SDGs into national programmes.

Policy transfer and NPG: scoping out communicative rationality

The policy transfer process is about multiple actors being guided by a collaborative form of rationality in their decision-making process. The policy outcome and its implementation are then parts of a joint product. This definition fits quite comfortably with the principles of the NPG paradigm of a pluralist state where multiple interdependent actors contribute to the design and delivery of policymaking and public services.

NPG, policy transfer, and the SSC fit

There is, however, very little systematic analysis of how an SSC type of policy transfer can take place within the NPG context. The NPG paradigm of a pluralist state (Osborne, 2010) is consistent with collaborative rationality (Healey, 1992). The NPG, an offshoot of the new public management (NPM) theory, is rooted firmly within institutional and collaborative governance. It posits both a plural state, where multiple interdependent actors contribute to the design of policies and delivery of public services, and a pluralist state, where multiple processes inform the policymaking system (Osborne, 2010). Thus, it calls for a collaborative rather than a prescriptive state. The term "collaborative state" is used to denote the institutionalized channels which enables communicative rationality thinking, making planning processes more pluralistic.[3]

A collaborative state is positioned closer to a liberal, open economy model than its prescriptive counterpart based on narrow scientific rationalism (Healey, 1992). Communicative rationality recognizes that planning is a complex process, requiring discussion and experimentation by a diverse set of state and non-state actors, where knowledge is created through scientific reasoning, and also through communication, exchanging perceptions, and understanding (ibid.).

It is not necessary that communicative rationality be internal to the state only. It can be international if there are common goals and aspirations. In such cases, it is feasible to contemplate a nation-to-nation collaboration aimed at the same outcome through discussion between state and non-state actors in the two states (or more).

Achieving SDGs represents such a common goal. In our case, Bangladesh and Peru are seeking to benefit from SSC by implementing SDG tracking processes through collaborative stakeholder consultations, using Bangladesh's SDG tracker as the primary initiator of the dialogue. Healey (2010) called for communicative rationality to leverage the interplay of structure and agency. The Bangladesh-Peru agreement suggests that states in the South have become more collaborative as they try to create conditions that would accelerate better SDG spending in partnership with others.

Collaboration theory under NPG posits two principles: (i) collaborative advantage, created through collective efforts or communicative rationality guiding decision-making; and (ii) collaborative inertia, created by a negative attitude towards change or the traditional, narrow, rational decision-making process (Huxham & Vangen, 2005).

Policy transfer is communicative rationality directing collaborative behaviour by different actors. It is about taking advantage of opportunities to improve performance by learning from a stock of knowledge from particular policy measures that have fared well in other contexts (Marsh & Sharman, 2009). Analysis of policy transfer is about "emulation and lessons drawing in which knowledge about policies, administrative arrangements, institutions etc. in one time and/or space is used in the development of policies, administrative arrangements and institutions in another time and/or space" (Dolowitz & Marsh, 1996, 344).

A government's decision to collaborate for the purpose of learning from another government follows the same logic. Selective use of evidence derived from policy lessons from another government can help to build credibility of an idea or legitimacy of a decision. An international consensus, such as the SDGs, can also offer an uncontroversial window of opportunity for SSC. The question is what form policy transfer will take once a government has decided to collaborate with another government.

Policy transfer is generally carried out using four mechanisms – learning, competition, coercion, and mimicry (Box 3.1). Their relevance and usage vary according to the purpose of policy transfer. For example, "learning is the main mechanism identified in the [policy] transfer literature, while other mechanisms receive more attention in the diffusion literature" (Marsh & Sharman, 2009, 271). While the policy transfer literature focuses on decision-making dynamics highlighting the role of agency, policy diffusion stresses the structure or the established

---

Box 3.1: Policy Transfer Mechanisms

- *Learning* implies a 'rational' decision by a government to emulate foreign success policy stories to improve functional efficiency.
- *Competition* is about the idea of a 'race to the bottom' whereby policies converge on a common low denominator.
- *Mimicry*, also known as emulation or socialization, is about copying foreign models in terms of symbolic or normative factors, rather than a technical or rational concern with functional efficiency.
- *Coercion* originates from powerful states or international organizations such as the International Monetary Fund (IMF) and World Bank via the conditions attached to their lending.

Source: Marsh and Sharman (2009)

frameworks through which new policies and lessons are to be diffused. Dolowitz and Marsh (2000) argued that policy transfer does not necessarily lead to policy diffusion, because policy transfer is a continuous variable. Marsh and Sharman (2009) recommend combining policy transfer and diffusion in order to explain the causality of policy change. In this chapter, policy change is viewed as primarily being driven by policy transfer as a result of SSC, which allows actors to learn from each other.

However, policy transfer and diffusion mechanisms, although analytically distinct, are not necessarily mutually exclusive. In other words, "how bounded does rationality have to be before learning becomes unreflective mimicry" (Marsh & Sharman, 2009, 272).

The reason for the bilateral SDG tracking agreement seems to focus on the learning mechanism using a process of communicative rationality. Multiple actors and processes contributed to joint decision-making to transfer successful policies and practices. Focusing on the learning aspect helps us to understand why policy transfer is taking place in the Global South in the first place.

The success and failure of policy transfer efforts can be viewed from two perspectives – (a) the extent to which policy transfer achieves the aims set by a government when it engaged in the transfer and (b) when key actors engaged in the process perceive policy transfer as a success (Dolowitz & Marsh, 2000). The latter is a key principle of the incrementalism approach where an agreement on policy itself is a test of a good policy (Lindblom, 1979).

Bangladesh-Peru collaboration: a policy transfer paradox

Marsh and Sharman (2009, 281) argued that "developing states legitimize themselves by mimicking developed states, rather than vice versa." Policy transfer in the context of SSC represent a paradox in this sense.

Table 3.1 presents some key development indicators for Bangladesh and Peru. Peru is an upper-middle-income country with a per capita income four times that of Bangladesh and a higher Human Development Index (HDI) rating. Bangladesh is a low-middle-income country albeit with a faster economic growth rate, but with a relatively higher poverty rate, and lower life expectancy and literacy rates. Nevertheless, the lower differential in the human development indicators relative to the difference in per capita income suggests that there is greater income inequality in Peru than in Bangladesh. This is borne out by the Gini coefficient estimates of 43.6 for Peru and 32.4 for Bangladesh from 2016. Peru has a better e-government ranking than Bangladesh, indicating the Peruvian government's greater success in using ICT tools to improve service delivery to citizens.

These facts may suggest that the direction of policy transfer should hypothetically be from Peru to Bangladesh, and the actual direction of the transfer (from Bangladesh to Peru) may be thought of as a paradox. However, the policy transfer direction can be partly explained by the strong emphasis placed by the government of Bangladesh on e-governance mechanisms (as noted by UN's e-Government Development Index report 2018).[4] E-governance has

Table 3.1 Comparison of Bangladesh-Peru income and development status

Indicators	Bangladesh	Peru
GDP growth (%) (2017)	7.1	2.5
GNI per capita, US$ (2017)	1,470 (low-middle-income)	5,960 (upper-middle-income)
Gini coefficient (2016)	32.4	43.6
Poverty rate (2016)	24.3	20.7
Life expectancy, in years (2016)	72.5	75
Literacy rate (2016)	72.8	94.2
Human Development Index (HDI) rank (out of 189 countries in 2018)	136	89
e-government rank (out of 193 countries in 2018)	115	77

Source: Compiled from various sources (World Bank's World Development Indicators, United Nations Development Programme's (UNDP) Human Development Index Trends 1990–2017, and UN Department of Economic and Social Affairs (UNDESA), 2018)

been recognized as one of the highest priorities of the administration including the PMO and cabinet division (UNDESA, 2016). Although Bangladesh (115) ranked below Peru (77) in e-government ranking, it was placed in the same category of countries going through a high-level of online service delivery transformation such as Canada, the United States, the United Kingdom, Estonia, India, Singapore, South Korea, and other high-income, developed countries (UNDESA, 2018).

The other reason for the apparent paradox may be that Bangladesh launched its SDG tracker earlier than Peru and had built internal capacity and skills to advance on it. According to interviews with Bangladeshi government officials, many countries richer than Bangladesh are emulating many of its successful development initiatives. The government of Peru will be drawing upon Bangladesh's SDG tracker's design and implementation process. For Bangladesh, emulation of its development efforts by other and often richer and more developed countries, bestows it with credibility helping it to gain a more favourable image and further legitimacy for its other ongoing and future initiatives.

Bangladesh-Peru MOU on SDG tracking: process and outcomes

The Bangladesh-Peru MOU was signed during the Second UN World Data Forum in Dubai in October 2018. Representatives from a2i of the Bangladesh government and *Instituto Nacional de Estadística e Informática* or the National Institute of Statistics and Informatics (INEI) from the Peru government were the signatories.[5] The MOU formalized sharing of technical knowledge and solutions on design and use of SDG trackers between the two countries. In order to obtain information about this SSC MOU process and the learnings from it, ten open-ended questions were communicated via e-mail to the MOU participants prior to interviews. The responses were finalized over e-mail communication ensuring high accuracy in the interpretation of interview findings. The interview questions dealt with five specific issues:

- SSC as a learning type of policy transfer mechanism
- SSC's timing and leveraging the "window of opportunity"
- Internal and external procedures formalizing a policy transfer process
- Post-MOU status of the Bangladesh-Peru partnership (as of March 31, 2019)
- The role of international agencies in building momentum and catalysing SSC

Learning through South-South policy transfer

Officials from Peru and Bangladesh agreed that SSC works when there is a willingness to overcome common problems through joint, collaborative efforts. The Bangladesh-Peru SSC for statistical learning is allowing the two governments to share their technical knowledge on how they are tracking SDGs. According to CIFF representatives, SSC can be a potential tool to track SDG progress because it enables all parties involved to learn about each other's strengths and weaknesses, creating a smarter solution than what would have been achieved individually.

INEI is learning how a2i designed and implemented the Bangladesh SDG tracker system. This has sped up the process for the launch of Peru's own SDG tracker later in 2019. For Bangladesh, learning about how Peru is dealing with its challenges as an upper-middle-income economy is helping the country to further tailor its SDG tracker system, as Bangladesh gets closer to that threshold. This SSC can be regarded as a voluntary form of the policy transfer learning mechanism encouraging Bangladeshi and Peruvian governments to collaborate for the achievement of common and individually advantageous goals.

Timing of SSC

The Bangladesh and Peru MOU fulfils a need to explore successful methods and solutions to strengthen the digitization of data collection practices. While a2i has earned a lot of credibility in terms of implementing the e-government agenda of the Bangladesh government, INEI has technical expertise on how to track SDG targets using surveys measuring the living and poverty conditions, and maternal and child health, which have provided information that have contributed to significant poverty reduction in Peru. Thus, both government agencies found a common interest in cooperating with each other in order to learn about each other's successful practices.

CIFF played an important role in triggering and formalizing the SSC collaborative process. In October 2017, a former minister of Peru, Ms. Ariela Luna, visited Bangladesh. CIFF helped set up a meeting between her and the a2i. Ms. Luna was surprised to learn that Bangladesh had already launched its SDG tracker at a side event organized during the 72nd United Nations General Assembly (UNGA) in New York in September 2017, while Peru was still in the design phase with its own tracker. It was clear to both her and a2i that both governments could learn from each other's SDG data tracking system designs, as they were facing common problems in making the tracker useful for decision-making by the government and donor community.

Following her visit, Ms. Luna introduced INEI and a2i over e-mail and CIFF assumed the role of a moderator in connecting both governments. CIFF was interested because such collaboration could lead to better tracking of nutrition-related aid investments across developing countries, which could help to improve achievement of many of the SDG targets and indicators. For CIFF, an SSC type of collaborative effort could help to design evidence-based nutrition service delivery programmes as the governments engaged in a collaborative learning process.

Formalizing policy transfer

The procedure for approving an MOU varies between different governments and agencies. It took approximately six months (from January to June 2018) for INEI and a2i to agree on a final version of the MOU before it could be sent off for internal validation and approval in both countries, which took another two months (July – August 2018). Over the first six months, CIFF organized 19 bilateral discussions between the two government entities. The interviewed respondents indicated that their organizations had strong relationships with their respective foreign ministries, which was useful in accelerating the internal validation and approval process.

For Bangladesh, first, the project director of a2i needed to sign off on the MOU. Then the internal procedure required getting approval from an inter-ministerial meeting organized by a2i, an approval note from the secretary of the ICT Division, and a consent letter from the Law and Justice Division under the Ministry of Law, Justice, and Parliamentary Affairs. And the Ministry of Foreign Affairs had to vet the MOU.

For Peru, although the procedure was much simpler, it took the same amount of time. INEI has the direct authority to sign agreements relating to technical assistance on statistical issues. Since the a2i-INEI collaboration is linked to monitoring of SDG indicators, the MOU was well within the scope of INEI's mandate, meaning that the INEI could subscribe to it directly. Nevertheless, the process took two months because the final MOU needed to be reviewed by Peru's Foreign Ministry and its chancellery office in New Delhi (as Peru does not have an embassy in Dhaka), before getting a sign-off from the head of the INEI. A further delay occurred because under Bangladeshi law, two witnesses are required to sign the MOU in order for it to become fully authorized, which was not the case for Peru. The INEI then had to decide internally who their second witness would be and whether there were any legal incompatibilities with this amendment.

There was a consensus among all respondents that the most important driving factor in the drafting of the MOU and its approval process was the willingness of both government agencies to learn from each other, and the persistence of CIFF in leading the SSC process. There were repeated interactions (19 formal and more than 25 informal meetings) over the six months before the actual signing of the agreement in Dubai in October 2018. CIFF's support was instrumental here because there were significant linguistic, cultural, and time difference barriers that needed to be taken into account for the discussions to happen on schedule.

CIFF supported travel, accommodation, and expenses for both government agencies for the MOU signing event. The Second UN World Data Forum in October 2018 presented a golden and highly visible opportunity for the formal signing of the Bangladesh-Peru MOU. INEI and a2i had from the beginning recognized the partnership as mutually beneficial and a good match in terms of their respective needs and strengths. There was also a mutual recognition of achieving a common goal – tracking SDG progress for better decision-making in SDG investments. This purpose served as a stimulant in generating momentum behind the partnership and in taking the steps to reach formal agreement.

One of the main achievements of the Cape Town Global Action Plan (CTGAP) in 2017 (a UN instrument signed during the Second UN World Data Forum) was highlighting the need for promoting partnerships to produce data on SDGs for improving effectiveness of aid and public money (Dubai Declaration, 2018). The MOU signed between a2i and INEI represented a strong step in this direction. The interviewees indicated that developing countries such as Laos, Ethiopia, and Afghanistan expressed an interest in joining the Bangladesh-Peru collaborative effort. This will certainly make the existing SSC between the two countries more pluralistic as more actors and processes are integrated into its framework.

Post-MOU signing status

A draft roadmap for collaboration was submitted by INEI in January 2019 and is being currently discussed with a2i. It outlines key activities such as country visits by each government, assessing the baseline for measuring progress in achieving SDGs, methodological improvements, and updating the SDG monitoring system for smarter SDG investments. In addition, there is going to be further cooperation with other government agencies in Bangladesh outside of a2i, such as with the BBS and the GED as well as the SSC cell under the Ministry of Finance. This will ensure that the technological and statistical aspects of the collaboration are properly institutionalized for scaling up the successful lessons, perhaps in other countries.

The INEI and a2i will carry out a peer evaluation to determine the current development status of the SDG indicators in both countries. The 2019 roadmap includes the development of methodological proposals for more complex SDG indicators. For example, the technical team of INEI plans to address 10% of SDG level II indicators. Furthermore, INEI is aiming to develop tentative methodological proposals for at least five indicators related to poverty and nutrition. This will need a more coordinated effort to allow for joint solutions for a greater exchange of technical knowledge and information on how to improve tracking of human development–related SDGs.

Bangladesh and Peru are now finalizing the roadmap. One of the first activities will be to assess country progress for the SDG indicators, which includes reviewing data sources and methodologies for each data tier. This exercise will help to establish a learning mechanism or bridge between the two countries by identifying specific targets they can collaborate on. A report is due in December 2019 which will detail the progress and achievement of the activities proposed in the framework of the MOU. Monitoring and sharing of lessons learned directly and indirectly are expected to contribute to new ways for sharpening SDG-17's concept and design.

Role of international development agencies

UN agencies could play a role in scaling up this SSC effort by facilitating the participation of countries like Laos, Ethiopia, and Afghanistan. International donor agencies like DFAIT, World Bank, UNOSSC, and United Nations Development Programme (UNDP) as well as philanthropic entities like CIFF have a role to play in this process. They can help to channel SSC initiatives in creating policy transfer learning mechanisms for encouraging evidence-based actions, by matching countries that are facing common problems.

CIFF has so far been successful in helping the Bangladesh-Peru discussions. This partnership is the first of its kind in terms of content (SDG tracking) and disparate geography (the countries are separated by a distance of about 18,000 km). If CIFF can continue to successfully facilitate the Bangladesh-Peru collaboration on SDG tracker learning, then others may see the value in prioritizing SSC for improving SDG spending effectiveness through building SDG tracking systems.

The international custodians of individual SDG indicators are also tracking SDG progress using data for specific indicators. For example, UNICEF, as the custodian or co-custodian for 17 SDG indicators, has created a data portal for these indicators. The Sustainable Development Solutions Network (SDSN) and Bertelsmann Stiftung have used internationally comparable data to produce the "SDG Index and SDG Dashboards" reports (Sachs et al., 2018).

The latest SDG Index and Dashboards Report 2018 by the UN Sustainable Development Solutions Network (UN-SDSN) ranked Bangladesh 111 and Peru 66 out of 156 countries. Given that the latter is learning from Bangladesh's "development mirror" and has yet to launch its SDG tracker, the rankings could have been made based on non–SDG tracker data or data gathered from non – credible sources. It is necessary to harmonize data between different governments and the existing platforms on SDG trackers because this could enable developing countries to learn from each other. SSC would appear to be the best vehicle to achieve the latter. International development agencies could help construct this South-South bridge by connecting different data portals and reporting processes on specific indicators, targets, and goals, and bringing governments together to work collaboratively towards achieving SDGs.

SSC for better SDG spending

The world will spend US$5–7 trillion on the SDGs. SSC can help to spend this money more smartly. The SDGs themselves recognize the potential for SSC-type partnerships so that countries can collaborate and learn from each other. SDG-17 is about the global community's role in promoting such collaborative forms to create partnership between countries in the Global South. There is evidence that SSC can do "phenomenal" good for the world because it will entail high benefits by enabling countries to share knowledge on mutually agreed terms (Lomborg, 2018).

This chapter has provided an analysis of a formal SSC type of policy transfer from a low-middle-income to an upper-middle-income country. Investing in SSC can improve the process of designing solutions that work well in other countries and can be tailored to meet individual country needs. The precursor of SDGs – that is, the MDGs – has created a fertile ground for such learning and knowledge exchange opportunities.

Global coordination

The UNOSSC's purpose is to mainstream SSC across the UN system by supporting developing countries' efforts to manage, design, and implement SSC policies and initiatives through identification, sharing, and transfer of successful development solutions. It manages the UN fund for SSC and the Pérez-Guerrero Trust Fund for Economic and Technical Cooperation among developing countries. However, there is little synergy between UNOSSC and other international data custodians like UNICEF, World Bank, and UNDP. The challenge is to establish a direct bridge between such data custodians and the UNOSSC in order to leverage the use of data for policy decisions and programme design purposes.

By engaging in a deliberative learning process on SDG trackers and portals across the globe, the UNOSSC could help to connect countries in solving common problems. The Bangladesh-Peru SSC in SDG tracking could serve as a pilot for other similar efforts. The a2i has been working with UNOSSC since 2016, and they have partnered in creating the South-South Network for Public Service Innovation, or SSN4PSI. The SSN4PSI is headed by the director of UNOSSC, and its secretariat is run by a2i. The UNOSSC and SSN4PSI recently published a report on SSC taking place in the Global South (UNOSSC & SSN4PSI, 2019). This type of national-global collaboration can be helpful in motivating other developing countries to join SSC initiatives for learning about Southern innovations in increasing development effectiveness.

National dynamics

The Bangladesh-Peru experience suggests that successful SSC requires repeated interactions between the responsible agencies for building trust and confidence in the process and with each other. Repeated interactions can stimulate learning and knowledge exchange, a key characteristic of collaboration theory. Modest aims at the beginning are preferable, but the definition of a longer-term (agreed upon) outcome is necessary. When the relevant national actors see the value in the policy transfer process through the establishment of a learning mechanism, they will be more willing to form a formal coalition.

A formal but non-legal agreement like an MOU is essential to make relations visible and real and to boost the enthusiasm of all involved in the process. For example, signing of the SDG tracking MOU by INEI and a2i officials gave it and those responsible high visibility in addition to demonstrating the value of the work being undertaken. Finally, the role of change agents like CIFF, Ms. Ariel Luna, and representatives from a2i and INEI are critical.

In terms of prospects for SSC in the area of policy transfer, the principal challenges relate to the political feasibility of the policy change effort and the delineation of the roles of policy agents and entrepreneurs. For example, will groups who yield high influence in the area under consideration intervene to disturb the process of collaborative learning? Will actors keen on SSC have the capacity and authority to overcome such structural barriers? More research also needs to be done to study policy transfer's operational, financial, and regulatory frameworks in relation to its broader public policy and political economy implications. Facilitation by international agencies and philanthropic organizations (like CIFF) can help bring governments together, as these organizations are perceived to be more neutral than the bilateral participants themselves.

Notes

1 We interviewed the following representatives: deputy chief of Peru's National Institute of Statistics and Informatics (or INEI), Dr. Anibal Sanchez Aguilar; policy advisor of Bangladesh's Access to Information (a2i), Mr. Anir Chowdhury; and the director of Child Health and Development at CIFF, Dr. Sufia Askari. Ms. Sadia Afrose Shompa, a project assistant at a2i and Ms. Celia Carbajosa, a consultant at CIFF, provided key support in finalizing interviews and providing materials for research.
2 The a2i is a special programme of the government of Bangladesh aimed at digitizing service delivery through innovations which simplify the processes. For more information, please visit: https://a2i.gov.bd/ (accessed on March 31, 2019).
3 See Healey (1992) and Osborne (2010) for a discussion on collaborative rationality and NPG, respectively.
4 The UN Department of Economic and Social Affairs (UNDESA) prepares the e-GDI to assess the performance of UN member states in using technology for improving information and service delivery to citizens, as well as increasing citizens engagement through e-participation.
5 The event was coordinated and funded by CIFF, UNOSSC, UN Statistics Division, and other UN agencies.

References

a2i. (2018) SDG tracker: Bangladesh's development mirror. Available at: https://a2i.gov.bd/wp-content/uploads/2018/01/SDG-Tracker.pdf [Accessed 31 March 2019].

Askari, S., & Luna, A. (2019) Data partnership is groundbreaking win for South-South cooperation. Available at: https://ciff.org/news/ground-breaking-win-south-south-cooperation/ [Accessed 31 March 2019].

Crosby, B., Bryson, J., & Stone, M. (2010) Leading across frontiers: How visionary leaders integrate people, processes, structures and resources, in Osborne, S. (ed.), *The New Public Governance? Emerging Perspectives on the Theory and Practice of Public Governance*. New York, Routledge, pp. 200–222.

Dolowitz, D., & Marsh, D. (1996) Who learns what from whom: A review of the policy transfer literature, *Political Studies*, 44(2), pp. 343–357. https://doi.org/10.1111/j.1467-9248.1996.tb00334.x

Dolowitz, D., & Marsh, D. (2000) Learning from abroad: The role of policy transfer in contemporary policy-making, *Governance*, 13(1), pp. 5–23. https://doi.org/10.1111/0952-1895.00121

Dubai Declaration. (2018) Dubai declaration supporting the implementation of the Cape Town global action plan for sustainable development data. Dubai. Available at: https://undataforum.org/WorldDataForum/wp-content/uploads/2018/10/Dubai_Declaration_on_CTGAP_24_ctober-2018_online.pdf

GED. (2017) Bangladesh development journey with SDGs: Prepared for Bangladesh delegation to 72nd UNGA session 2017. Dhaka. Available at: https://info.undp.org/docs/pdc/Documents/BGD/UNGA_Booklet_2017.pdf

Healey, P. (1992) Planning through debate: The communicative turn in planning theory, *The Town Planning Review*, 63(2), 143–162. https://doi.org/10.2307/40113141

Healey, P. (2010) *Making Better Places the Planning Project in the Twenty-First Century*. Basingstoke, Palgrave Macmillan.

Huxham, C., & Vangen, S. (2005) *Managing to Collaborate: The Theory and Practice of Collaborative Advantage*. New York, Routledge.

Lindblom, C. E. (1979) Still muddling, not yet through, *Public Administration Review*, 39(6), p. 517. https://doi.org/10.2307/976178

Lomborg, B. (ed.) (2018) *Prioritizing Development: A Cost Benefit Analysis of the United Nations' Sustainable Development Goals*. Cambridge, Cambridge University Press.

Marsh, D., & Sharman, J. C. (2009) Policy diffusion and policy transfer, *Policy Studies*, 30(3), pp. 269–288. https://doi.org/10.1080/01442870902863851

Osborne, S. (2010) *The New Public Governance? Emerging Perspectives on the Theory and Practice of Public Governance*. New York, Routledge.

Rahman, S. S., & Baranyi, S. (2018) Beyond binaries: Constructing new development partnerships with middle-income countries, *Canadian Journal of Development Studies/ Revue Canadienne d'études Du Développement*, 39(2), 252–269. https://doi.org/10.1080/02255189.2017.1394275

Sachs, J., Schmidt-Traub, G., Kroll, C., Lafortune, G., & Fuller, G. (2018) *SDG Index and Dashboards Report 2018 Global Responsibilities: Implementing the Goals*. New York, PICA.

UNDESA. (2016) *United Nations E-Government Survey 2016: E-Government in Support of Sustainable Development*. New York, United Nations. https://doi.org/978-92-1-123205-9

UNDESA. (2018) *United Nations E-Government Survey 2018: Gearing E-Government to Support Transformation Towards Sustainable And Resilient Socities*. New York, United Nations.

UNDP. (2013) *Human Development Report 2013: The Rise of the South: Human Progress in a Diverse World*. New York, United Nations.

UNDP. (2016) *Human Development Report 2016: Human Development for Everyone*. New York, United Nations.

UNOSSC. (2018) *Good Practices in South-South and Triangular Cooperation for Sustainable Development – Volume 2*. New York, United Nations. Available at: https://drive.google.com/file/d/1pMjBpMwMDPPBD_EKtnp0zL9TxsQ3tmGT/view

UNOSSC & SSN4PSI. (2019) *The South-South Matchmaker: South-South Network for Public Service Innovation 2017–2018 Best Practices*. Bangkok, UNOSSC & SSN4PSI. Available at: https://a2i.gov.bd/wp-content/uploads/2019/03/The-South-South-Matchmaker.pdf

Wagner, L. (2018) Policy brief: Getting to 2030: Tracking SDG indicators for evidence of implementation progress. Available at: http://sdg.iisd.org/commentary/policy-briefs/getting-to-2030-tracking-sdg-indicators-for-evidence-of-implementation-progress/ [Accessed 31 March 2019].

Part III

South Asian landscape

Moving from patronage to participation

Part III

South Asian landscape

Moving from patronage to participation

4 Society for Elimination of Rural Poverty (SERP)

An Indian experience of rapid poverty reduction through women's empowerment

Shoaib Sultan Khan

Impressed by the success of the Aga Khan Rural Support Programme (AKRSP), UNDP invited the General Manager of AKRSP, Shoaib Sultan Khan (SSK), to replicate his approach across South Asia. The government of Andhra Pradesh in India was the first state government to demonstrate his approach. At the outset, he was asked to state his conditions for successful programme implementation. SSK replied that if communities want to overcome poverty and they are willing to organize, they can rise above the poverty level. The next question asked was about the scale of the operation he had in mind. SSK said that he would start from a village. In due course the Society for Elimination of Rural Poverty (SERP) was set up by the government of India to continue Khan's work. By now SERP has been able to lift 11.55 million households above poverty level by organizing and training women's self-help groups (SHGs) and providing them credit for microenterprise development.

Keywords

Poverty alleviation, community development, social entrepreneurship, seed capital, self-help groups, Andhra Pradesh, India

Background[1]

In 1992 when the foreign secretary of Pakistan rung me up late at night in Gilgit to inform me that the prime minister of Pakistan had nominated me on the Independent South Asian Commission on Poverty Alleviation, I had no idea what I was getting myself into. It turned out to be one of my most productive and enjoyable assignments and far reaching in determining my future destiny.

The South Asian Association for Regional Cooperation (SAARC) summit held in Colombo in 1991 established an Independent South Asian Commission on Poverty Alleviation (ISACPA). The Commission's recommendations led to the Dhaka Declaration, which sought to eradicate poverty by 2002. In 1993, after the endorsement of the commission's overarching recommendation by the SAARC heads of state that social mobilization should be the centrepiece for all poverty alleviation strategies of South Asian States, another event of great significance took place in Annapolis, United States, where the report was being discussed in a workshop jointly organized by the World Bank and the SAARC secretariat. By coincidence, Mr. Henning Karcher, Chief of UNDP's Regional Bureau for Asia and the Pacific (RBAP), happened to be at the workshop and a month later invited me to present the concept of social mobilization espoused by the Poverty Commission at the cluster meeting of the resident representatives of UNDP at Kathmandu. The idea of a South Asia Poverty Alleviation Programme (SAPAP) took birth at that meeting and UNDP asked me to help in operationalizing the concept of social mobilization in South Asian countries.

It was a daunting challenge, but I accepted it because I had full trust and confidence in my friends and fellow commissioners of the Poverty Commission. Foremost among them was Mr. Venugopal, secretary to the prime minister of India, without whose support and guidance the Indian component of the SAPAP would never have materialized. As they say, nothing grows under the Bunyan tree, and UNDP headquarters felt that the SAPAP would not thrive in Pakistan in the presence of AKRSP. However, they had to give in to the pressure by the Country Office, and the Pakistan component of the SAPAP has a place of its own in the family of Rural Support Programmes in Pakistan in reaching the poorest of the poor. The role of the UNDP Country Offices in promoting, implementing, and guiding the country components was most critical and vital in the progress of the SAPAP. The Country Offices' support to me personally was overwhelming, making my work much easier in each country.

The SAPAP, I was told, was something new and different in UNDP. But the way rules and procedures were interpreted in favour of the SAPAP was what made its implementation possible. At headquarters, the SAPAP always received the patronage and support of Dr. Nay Htun, Assistant Secretary General and head of UNDP Bureau of Asia and the Pacific. Mr. Henning Karcher was always looked upon as a guardian angel and Mohammad and Saraswathi as great well-wishers. The SAPAP was also most fortunate in getting the personal support of secretary general SAARC Mr. Naeem U. Hasan and SAARC secretariat ably given by Director Chowdhry.

However, it is the villagers of South Asia in Bangladesh, India, the Maldives, Nepal, Pakistan, and Sri Lanka who gave substance and reality to the SAPAP. The strategy of social mobilization espoused by the SAPAP was based on the assumption that people, especially the poor, are willing to do many things themselves and that there is tremendous potential lying dormant to be harnessed. If the rural poor would have nullified this assumption, the SAPAP would have had no legs to stand on. This is the main distinguishing feature of the SAPAP approach. It did not approach the rural poor with a preconceived package of services. It was not a service delivery programme. It helped people to help themselves. This is how it did it.

The SAPAP demonstrated the importance of support organizations in initiating the process of social mobilization in order to harness people's potential to help themselves. Without a dedicated, committed, and well-trained team, constituting a

Figure 4.1 First Dialogue – Sherqila (1982)

support organization, social mobilization is not possible. The team itself requires long-term support if it is to respond to the potential that the social mobilization process unleashes in the poor for rising above the level of subsistence. Social mobilization is a movement which cannot be switched on and off depending on the availability of resources. It takes time to build up momentum, which, if it is lost due to lack of access to resources in a timely manner, is difficult to rekindle, as the programmes are likely to lose credibility in the eyes of both the communities and the support organization.

The intent: first there was the word

As mentioned earlier, the recommendations of the ISACPA led to the Dhaka Declaration with the goal of eradicating poverty by 2002. The India component of the project, titled "Institutional Development at the Grassroots for Poverty Alleviation," started in 1994 with a focus on enabling the voiceless to articulate their needs. The project sought to alleviate poverty by handing over the organization and management of developmental affairs to the village communities. The project used thrift as an entry point. Programmes were largely implemented through local NGOs and grassroots people's institutions.

The political intent of SAARC leaders turned into a reality because a well-tested model for poverty alleviation existed in the northern areas of Pakistan in the form of the Aga Khan Rural Support Programme (AKRSP). UNDP subsequently came on board to scale up this approach to South Asia level under a new initiative, the SAPAP.

Vision and strategy: support organization and social entrepreneurship

The origins of the Andhra Pradesh District Poverty Initiative Project (APDPIP), and the later Andhra Pradesh Rural Poverty Reduction Project (APRPRP), lay in Indian women's savings-and-loan associations and the UNDP supported by the SAPAP, which grew out of the Dhaka Declaration. These approaches were also based on experience elsewhere in South Asia, including the Rural Support Programmes in Pakistan and the Self-Help Affinity Groups implemented in Karnataka, Andhra Pradesh, and Tamil Nadu by the NGOs MYRADA and PRADAN.

In India, during the decade prior to the SAPAP, there was increasing concern about the failure of anti-poverty programmes to reduce the number of rural poor. There were a number of problems with these programmes including poor targeting, weak management, lack of complementarities, and wasteful subsidies. The state's public spending on social development was low and declining. Social indicators were weak and below the national average. Some 30% of the population lived below the poverty line, and malnutrition in children up to 6 years old was about 30%. Notwithstanding some earlier experience in India with women's self-help groups and joint lending, there was limited availability of credit partly due to the high cost and risks of lending in rural areas to households with small borrowing needs and climatic variability. Line agencies were weak with a limited capacity for providing support and services for the poor. Agricultural productivity needed to be raised to enhance livelihoods. There had been a bank-funded economic restructuring project (Cr./Ln.49385/IN), but it had been unable to achieve direct coverage of the more vulnerable households. The government of Andhra Pradesh recognized the need for improved policies and the development of more effective institutions, as well as an approach that empowered poor rural communities, particularly women. In 1999, seeing the success of the SAPAP, the state embarked on a programme based on a 2020 vision strategy. Central to the strategy was the development of women's self-help groups (SHGs) that would be linked to commercial banks to increase the flow of, and reduce the cost of, credit. All line agencies were to adopt more responsive approaches to communities.

To meet this need the SAPAP had introduced a programme strategy inspired by the AKRSP model relying on the triad of social organization of the poor to be facilitated through "social guidance"'; skill development to nurture inherent capacities and potentialities to widen the basket of opportunities for the poor; and capital formation for improved quality of life and overcoming hunger, destitution, and deprivation on a sustainable basis through regular savings and judicious capital investments. The project initially covered 695 households in 20 Mandals[2] in three drought-prone districts of Andhra Pradesh – namely, Anantapur, Mahbubnagar, and Kurnool. As per the 1991 census, the population of the 20 Mandals covered was 824,113 and consisted of 156, 274 households. In 1995, the National Project Coordinator (NPC), much against my advice, expanded the project to the entirety of the three districts, but the coverage was reduced to 20 Mandals in 1996 by the new NPC on my advice as it was experienced that the area was too large

Table 4.1 Mandals covered in the pilot districts

1. Daulatabad	2. Bommaraspet	3. Kosgi	4. Peddakothapally
5. Pangal	6. Addakal	7. Mahanandi	8. Peapully
9. Orvakal	10. Midthur	11. Gadivemula	12. Dhone
13. Panyam	14. Peddavaduguru	15. Gandlapenta	16. Hindupur
17. Madakasira	18. Gudibanda	19. Bathalapalli	20. B. Samudram

for effective social mobilization. By May 1997 the project organized the poor into 2,395 self-help groups with 42,842 members in association with 18 partner voluntary organizations. The NGOs were supported through the project to provide social guidance to the poor.

The project also supported the gradual development of community resource persons at the Mandal (sub-district) level to support capacity building within the federated approach. Resource persons were recruited mainly from the more successful SHGs federated into village organizations forming Mandal Samakhya, who had demonstrated leadership and had the trust of their peers. Initially funded by SERP, these organizers were eventually paid for their services by the communities themselves. The programme received funding from UNDP under project RAS/94/500.

Dialogue with the partners – the first principle[3]

The cardinal pillar of the AKRSP approach was to get political, administrative. and community leadership of an area on board before reaching out to low-income communities. The same approach was followed during programme replication in Andhra Pradesh, India. The programme received unmitigated support, from the highest state officials to district collectors and Mandal officials from its inception in December 1994.

Before the advent of the project there were 232,000 women already organized in the three programme districts through Development of Women and Children in Rural Areas (DWACRA) and other programmes and NGOs. The first project NPC planned to add another 264,064 women to this figure. These interventions comprised only one input – namely, capacity building through a programme of training of SHG leaders for a duration of 15 days spread over a year, through partner NGOs. However, in the few groups visited by me, the need for responding to the development perceptions of the communities, which seemed to not be receiving the attention they deserved, became apparent. It was imperative to take up a canvas of activity large enough to influence policymakers and other agencies and viable enough for intensive operations beginning with capacity building of group/community leaders, capital generation through savings, formulation of an investment/development plan comprising portfolio of community perceptions, accessing of resources from financial and other institutions, linkages with government and other donor-supported programmes, and projects in the field of poverty alleviation. The new NPC, after extensive consultation and discussion, came to the conclusion that a sample of about 72,000 women organized in the three districts during the year 1996 could provide deep insight and also enable demonstration of trends in poverty alleviation. A logical framework was accordingly drawn up.

The new National Project Coordinator (NPC), Mr. K. Raju, requested me personally to administer a one-time grant of approximately US$100,000 to US$125,000 to partner NGOs as credit in 1996, in addition to inputs for the 72,000-person demonstration pilot. Such an investment in the NGOs would help in scaling up the results obtained from the demonstration pilot in the future. To give a practical shape to this decision, the NPC, along with programme managers Anantpur, Kurnool, and Mahboobnagar as well as Mr. Manohar, Dr. Fatima Ali Khan, and Ms. Vijayabharti, drew up a detailed work plan for the first quarter (January–March 1996) based on the logical framework prepared by the team earlier in January in consultation with Neera Burra of UNDP and myself.

When I asked UNDP for the money requested by Raju for giving credit, the UNDP New York Finance Manager Ms. Ingunde flatly turned down the request as the UN charter did not allow UNDP to give credit. On my retort to close the shop as these rural poor women were only earning Rs. 5 per day weeding fields of the land owners and there was no other way to help them but to give them capital to do things to generate income which they are capable of doing, Ingunde responded that she had not refused to give capital but she could not take it back by giving it as credit. This gave me the idea of taking grant money from UNDP and expensing it out of SAPAP books by converting it into a seed capital grant to SHGs with the directive not to spend it but convert it into a revolving fund managed by SHGs. The funds would be disbursed to members according to the requirements of their individual micro-investment plans on terms and conditions determined by the SHGs with the assurance that members taking seed capital would be eligible for future access to the revolving seed capital fund after returning the first amount. The initiation of seed capital as a revolving fund at the disposal of each SHG resulted in a manifold increase of the amount of available seed capital over a period of two to three years. This increase enabled the SHGs to access credit from commercial banks at a 12% rate of interest pledging their seed capital as collateral. In 2004, the chief minister decided to reimburse 75% of the interest amount charged by the banks on timely repayment of the loan amount by SHGs. This resulted in an exponential increase in the ability of the one million organized SHGs to access credit which totalled Rs. 100 billion from banks when I visited Andhra Pradesh in 2013.

Transforming government as a support organization

I had expressed the hope that within six months the nearly new team in India, under the dynamic leadership of NPC K. Raju, would be able to chalk out a clear plan of action. His team made a quantum jump much beyond my most optimistic estimates. Of course, their visit to AKRSP, according to them, helped them greatly in understanding the approach of harnessing the potential of the people to help themselves through social guidance by a support organization. My most pleasant surprise was the implementation of some of the plans which were having an incredible demonstration effect and building up programme credibility which nothing else could have done so quickly and so effectively. The men came to me requesting that they should be organized also. It reminded me of Gilgit, when women approached me asking why they were not being organized, when they were just as willing to fulfil the terms of partnership of the organization: capital accumulation through savings and upgrading of human skills by participation in human resource development.

During our dialogues in the field, Mr. Venugopal, on a field visit from Delhi, explained to the participants that RAS was a programme of the government of Andhra Pradesh, but it was not a programme to be implemented like other government programmes for the simple reason that the methods used to implement government programmes over the last 50 years did not make any demonstrable impact on poverty. The need was how to change those methods. There was need for a different paradigm and a different approach. Social mobilization is one such methodology which prepares people to receive whatever the government has planned for them. The RAS/96/600 project would help people build their capacity to take over responsibilities and management themselves. Conventionally the government did not talk to people or groups of women but prescribed an aspirin-type cure. Although RAS/96/600 was to be initiated in six Mandals, every effort would be made to cover the whole of the Kurnool district. Venugopal emphasized that RAS/96/600 would seek complete association with the district administration. I explained to the group that the vision of RAS/96/600 was not only to improve the situation of the poor in the Mandals or the districts it was operational in but to develop a cost-effective replicable strategy for reduction of poverty. I impressed upon the NGOs the need to keep this vision in view in their collaboration with RAS/96/600.

It is essential to carry out a feasibility study of the perceptions of the village women and men for preparation of micro-plans. The coordinator/CV should sit down with the groups and take up each identified need and work out in consultation with the community the beneficiary base, investment/credit required, cost-benefit of the venture, management and maintenance aspects, repayment schedule in the case of a line of credit from the seed capital fund, and overall impact on poverty alleviation of the intervention. The management and implementation capacity of the community, group, or individual should determine the amount of assistance or credit by RAS/96/600. Invariably a capacity-building programme of training and human resource development specific to the identified needs will have to be simultaneously initiated. Thus, the micro-plan should not be a shopping list but an operational plan with every conceivable detail as specified above. The district support team – that is, the project officer and the social organizer – with the assistance of district line agency specialists, should have the micro-plans, prepared by the Mandal support team in consultation with the

Figure 4.2 RGMVP Women's Group

people, vetted and recommended to the NPC for the allocation of resources either from the RAS/96/600 project seed capital or by accessing resources from other government departments or agencies, if available.

Moving NGOs from patronage to participation

During the dialogue with partner NGOs, it became apparent that the government's and NGOs' sense of urgency to deliver to the poor turned into a programme of patronage in the first NPC era. Out of 280 NGOs in Anantpur, 80 responded to the RAS/96/600 project invitation for partnerships. Of these, 20 were finally selected, and they engaged a total of 168 Karikartas (paid volunteers) and 26 coordinators. This worked out to 1 Karikarta per 20 groups and 1 coordinator for every 6.5 Karikartas. The partner NGOs were covering 36 out of 64 Mandals in the district, spread over 1,289 villages and 3,331 SHGs comprised of 51,008 women. District Coordinator Manohar agreed to intensively review the achievements of the partner NGOs and assess how successful they were in terms of output targets worked out in consultation with them. On the basis of the short field visit, it seemed very unlikely that the partner NGOs would be able to deliver. However, we agreed with Manohar that they should be given a chance and the position reviewed in six months' time. Meanwhile, we suggested to Manohar to have a fallback position by taking five Mandals and experimenting and developing different methodologies for coming up with a replicable strategy for the districts in the shortest possible time. In Mahbubnagar and Kurnool, the same strategy was recommended to the district coordinators and project professionals, Fatima, Murali, Premchandra, and Vijaya Bharti. This would entail direct interaction of the RAS/96/600 project staff with the villagers and a suggestion was made that a paid Mandal coordinator may be needed for each Mandal to assist the RAS/96/600 staff instead of the large number of paid volunteers (Karikartas) now being used by partner NGOs. The approach would be to foster a cadre of activists/idealists from amongst the communities; initially these activists would be volunteers like Vijaylaxmi of Ramachandrapuram but ultimately, they would be remunerated by the community itself.

We also felt that the NPC should reside in one of the project districts and have a management group of experts to enable provision of technical assistance to the organized groups. Initially Dr. Almas and Murali supplemented by a credit and enterprise development specialist were considered to be sufficient to help the NPC and the district teams to provide technical assistance in the fields of engineering, social sector and credit, and enterprise development. This created the possibility of demonstrating an alternative way of working through NGOs and helping them decide if they wanted to move away from a patronage-based approach to a participatory style of work.

Opening of the first shops:[4] carving the path for the snowball effect

As the work progressed, I was happy to see that the programme was gradually moving towards a holistic approach and the exclusive SGHs of women were being multiplied by a men's group and that both groups were feeling the need to start to meet as village organizations (VO) at least once a month. A representative from each group was inducted as the executive arm of the VO. I underscored the importance of keeping the executive arm subservient and accountable to the VO. A clear division of functions and responsibilities of the SHG, VO, SHG network at the village level and the Mandal level also seemed to be better understood and action was initiated by Raju to codify it. In one Mandal an informal bank even started functioning to access capital from credit institutions.

However, the most fascinating, exciting, and interesting part of the project was the discussion with group members on the investments in the organization plans prepared by them. To my great delight these turned out to be mainly a portfolio of opportunities identified and perceived by the members as within their capacity to implement. I have always believed that there is tremendous potential in people to improve their condition themselves, if only they can be helped to get over the handicaps which inhibit them from harnessing this potential. In village after village I visited, the members of the groups came up with clear ideas and had full confidence and determination to implement these ideas, if only they could also have access to capital and skills.

For the first time I felt nostalgic when I met the groups jointly either in the open under petromax lights or under mango groves or beneath a shamiana erected for the occasion. It was more like the holistic development meetings I was used to. The discipline and maturity of the groups and the standard of the interactions were most impressive in many places. Some groups were still struggling to get out of their earlier mould of simply thrift and savings groups. The discussion was always very candid and forthright. In one place when questioned as to why the member-ship of the groups only embraced half the households, the answer was that the others are waiting but already requests had been received for membership once the seed capital effects had been seen. In another village where seed capital was yet to be advanced, the answer was that non-members were not impressed by the performance of the groups over the past so many years in collaboration with the local NGO. However, there was no doubt in anybody's mind that once the VO and the groups were able to implement even a few initiatives included in the portfolio of opportunities, the non-members, especially the poor, would take advantage by getting into an organized fold.

The kind of opportunities members of groups identified embraced bore wells, submersible pump sets, electric motors, services like laundry, mini flour mills, dolomite crushers, seasonal businesses, petty businesses, bangle shops, provi-sions shops, bullock carts, and so on. But the majority opted for sheep rearing, goat rearing, heifer fattening, milch cattle, crossbred jersey cows, poultry, land lease, quarry lease, nursery development, and so on. In the field of capacity build-ing, they sought managerial, accountancy, livestock, horticultural, and agricul-tural training besides health and sanitation training. They were very conscious of the importance of family planning, the eradication of social evils such as child marriages and child labour, and, in a few places, I found Balika Sangams (youth or children's organizations) that put emphasis on education and members attend-ing night classes because of their compulsion to earn during the day. New avenues of income were also being explored such as agarbatti (incense) making and in some places even bidi (hand rolled cigarette) making.

When I quizzed members about their ability and capacity to undertake all these initiatives successfully, they stood their ground. I was joined by the secretary of the Department of Finance of the government of Andhra Pradesh, Ms. Sujatha Rao, in questioning them on how they were going to procure goods for their pro-posed supermarket or bunk (kiosk). Our concerns were unfounded because they showed us both the units fully stocked with goods and requested that we inaugu-rate the sales. This was the first day of the opening of the shops. They had taken full care of how they were going to look after their assets whether it was a cow or poultry or leased land or a business microenterprise.

A woman who had been a member of a group organized by an NGO four years ago, when asked what difference the organization had made to the improvement of her standard of living, answered 10%. When questioned about accelerating this process, she answered that she was an agricultural labourer and if given training in grafting, pruning, and other agricultural practices plus a small amount of capital as a loan, she would like to develop a plant nursery of her own. In addition, she wanted traditional birth attendant (TBA) training, which was a crying need of the village and would bring her a handsome income on deliveries. I asked what difference all this will make to her and in what time frame. Her prompt reply was that in six months' time, it would bring about 100% improvement in her standard of living. She needed no more than Rs. 5,000, including training costs and a loan of Rs. 3,000 to bring about this change.

In another dialogue when I asked the group if they had any ideas how they could better their lot, there was an animated response that "our heads are bursting with ideas." One of them, a vivacious and dynamic leader, said that she would like to do so many things because it takes her no more than a few days to plant her rain-fed one-acre plot with groundnut and the rest of the time she has to do wage labour and that, if her husband would allow her, she would do many things. I prompted, why don't you take your group members to get some sense hammered into his head? She said he would be very polite with them but clobber her as soon as the group left. The local NGO representative observed in English that this lady had had an open affair, and this was why her husband didn't trust her. Anyway, we sent for the husband, a mild-looking, harmless fellow, and asked him why he didn't allow his wife to put her ideas into practice. He said he had no objection to taking a loan for five ram-lambs, but he can't make out how they would be able to pay instalments monthly immediately after taking the loan when the income from the lambs would only come after six to seven months. We made a proposition of, if the instalments are deferred for six to seven months, would he agree to his wife's idea, and his answer was in the affirmative.

In another village, 47 educated unemployed youth came up with a proposal to install dolomite crushers at a total investment of Rs. 200,000 each, providing employment to seven educated youths and ten wage labourers. There was scope for many such crushers. I wondered if there was a more cost-effective way of generating employment. The wage labourers quarrying dolomite proposed loans to lease out ten-foot-wide quarry strips which would increase their daily income by Rs. 750. The total loan amount needed by a group of five workers was no more than Rs. 18,000. Another group of five requested a loan of Rs. 60,000 to use the dolomite dust for making blocks and earning Rs. 12,000 net per month. The agricultural group of 15 members wanted capital to lease land at Rs. 2,500 per acre for growing groundnut seed and expected to make a net income of Rs. 1,800 per acre.

Obulavaripalli Village, with 59 households, had formed a village organization with both men and women as members. Every household had participated in preparing the organization investment plan and every member had been included in the portfolio of opportunities requiring seed capital of Rs. 242,000 in addition to their own capital amounting to Rs. 140,000. But the VO also considered drinking water as a top priority and needed a subsidy of Rs. 60,000 to undertake the project with almost an equal contribution by the VO in the form of labour and local material. I wanted to know how strongly they felt about this project, so I made a proposition asking for volunteers to forego seed capital worth Rs. 60,000 to enable the RAS/96/600 project to respond to their request, as the total amount available was

Rs. 242,000. Almost ten members stood up foregoing their individual claims for the common good of the community. It did not leave any doubt in my mind about the cohesiveness and collective management capability of this VO.

What came out of these dialogues was that the assumption that rural people are willing to undertake many things themselves to help improve their situation was based on a very solid foundation, and if this potential was harnessed, it would make all the difference to their lives and was the most effective way to reduce poverty. I was also very happy to see that Raju had settled the thorny issue of collaboration and partnership with NGOs. Only those NGOs who fully subscribed to and believed in the strategy of social guidance being espoused by RAS/96/600 were now partners. When I look back, I feel that we learnt a great deal from our earlier experience of working with partner NGOs, and it helped us tremendously in forging the current norms of partnership.

In the review meeting with the team at Hindupur at the end of my sojourn by road from Hyderabad, many issues were discussed including the forthcoming midterm review and the additional requirement of resources to implement the VO plans. It was decided by the team to take the following action:

- Prepare 42 VO plans (2 per Mandal) immediately.
- Prepare projections of resource requirements for total coverage of 21 Mandals.
- Prepare projections for resource requirements for total coverage of the 180 Mandals in the three districts.
- Do an immediate estimate of total resources available in the RAS/96/600 project budget up to the end of 1998 (including reappropriations of savings) for assistance to communities.
- Undertake a survey and plan preparation of VOs strictly keeping in view the available resources. It would be counterproductive to raise hopes of villages for which resources had not been mobilized.

On the basis of the plans prepared, it was estimated that an average of Rs. 600,000 would be required per portfolio of opportunities and investments, of which 80% would be for returnable seed capital and the balance for collective activities. On average 50 plans per Mandal were likely to be prepared for total coverage requiring Rs. 30 million. Thus, the total requirement for the 21 intensive Mandals, currently the focus of the RAS/96/600 project, would be in the region of Rs. 600 million (60 crores). The importance of this demonstration can be judged by the fact that the Ministry of Rural Areas and Unemployment alone, according to Secretary Mr. G. P. Rao, had made an allocation of Rs. 75,000 crores for the next five years for poverty alleviation. He expressed keen interest in what the RAS/96/600 project was attempting to do in Andhra Pradesh and held out strong prospects of linkage with government programmes, in case of success.

Monitoring and learning: mirror, mirror on the wall . . .

In the meantime, I impressed upon Raju the urgency of recruiting a monitoring and evaluation expert in view of the forthcoming midterm evaluation (MTE) and proposed that the programme should hire a consultant to undertake impact studies, case studies, and cost-benefit analysis especially of the training programmes and seed capital interventions.

The MTE exercise was one of the most frustrating yet equally if not more euphoric weeks of my association with the SAPAP. I was stunned to hear from Henning Karcher that the leader of the recently completed evaluation team had expressed serious concerns about SAPAP, in her briefing in Delhi to Nay Htun and Henning, pertaining to the following:

- Non-targeting by the SAPAP of the poor
- Non-targeting of the aged and the destitute
- The gap between the leaders and the members
- Only 10–15% of the groups are viable in terms of strength
- Lack of convergence of services between groups and others such as government services, local councils, and so on
- The programme was expanding too fast. It should consolidate and deepen instead of furthering expansion, and groups should be linked to banks for obtaining credit.
- Seed capital is out of hand with little control since there is no relationship between loans and savings
- Lack of participatory planning in preparation of investment (micro) plans
- Dropouts due to fines
- Illiterates not able to access benefits
- Rs. 2,000–25,000 seed capital range contributing to intra/inter-village gulf between rich and the poor
- However, the programme is promising and must be continued

This all came as a shock to me because the team in Islamabad had the opportunity to raise these issues with me from October 25 to 31 and Ms. Mary Clark stayed up to November 2, 1997. But the only time they met me was at a social evening at my house, and the only issues they raised pertained to the future of the process of social mobilization: when will it end, the different country colours of the programme, the shape of support organizations with special reference to Bangladesh, and the macro-policy framework and monitoring. Not a word was said about the issues Ms. Mary Clark raised with Nay and Henning. Dr. Neela Mukherjee did not ask a single question and didn't seek any clarification from me.

On the contrary Shyam Khadka assured me of a very positive attitude of the evaluation team members towards the SAPAP on the basis of what they had seen on the ground in every country. He told me that the member from Denmark expressed the view that the SAPAP provided an opportunity for UNDP to move away from the supply and technical assistance mode to making a real impact at the grass-roots level by reducing poverty. Shyam felt that even Dr. Neela Mukherjee had come a long way, and her initially hostile attitude was becoming positive. Mary Clark told me that she had been writing about these things and was happy to see the translation of theory into practice. Thierry commented that he understood the inadequacy of the microcredit approach compared to the holistic approach adopted by the SAPAP. All the team members had nothing but good things to say and praise for whatever they saw on the ground and how inspiring the experience was and what a learning exercise it had been

In the backdrop of these comments Henning made a field visit to India. It was no less than providential. What he saw between Bangalore and Hyderabad during the 600-kilometer journey in the districts of Anantapur, Kurnool, and Mahbubnagar left him with no doubt about the soundness and viability of the strategy adopted by the SAPAP and the impact it had made in a short period of two to three years on the poor households. The programme included visits to many places and villages where we interacted with ordinary members of organizations, networks of SHGs at the village and Mandal levels, and groups of activists and partner NGO staff. Besides meetings and discussions with the programme staff at Anantapur, Kurnool, and Mahbubnagar, Henning also met a large number of district- and state-level government officials including the chief secretary and secretaries of planning and rural development. At every place Henning raised all the issues identified by the evaluation team and got a completely different picture, which was also fully testified to by the Participatory Rural Appraisal (PRA) of the Anantapur district undertaken by one of the three research teams engaged by the evaluation team. In fact, in Kurnool we saw how a village community had taken steps to make itself self-reliant. Village Kalva with 330 households below the poverty line had organized itself into 20 SHGs in the last two years with the help of the SAPAP, encompassing 277 households. It came up with a portfolio of opportunities requiring Rs. 3 million. Having received half that amount from the SAPAP as seed capital, the village network of SHGs gave loans at a 24% rate of interest to 147 households for an array of activities ranging from lines of credit for Rs. 5,000 to 7,000 for agricultural inputs to leasing of land by landless women for sugarcane cultivation to microenterprise, such as, wayside restaurant, boutique (bunk), small flour mill, and building material business, as well as bullocks and horses for carts, and so on. The organization also helped set up a centre to help the disabled and the destitute and a centre for adolescent girls to learn about tailoring, embroidery, horticulture (kitchen gardens), the environment, nutrition, health, and hygiene.

The centre also helps in the prevention of early marriages. The SHGs were not only aware of family planning practices but are also advocating their adoption amongst members. But the most impressive achievement of Kalva Village Organization was their decision to hire a local graduate in commerce to act as the accountant of the Network at a monthly salary of Rs. 2,000 paid by the Network. This was the first step in achieving sustainability and independence from NGOs or the SAPAP.

In every village that Henning went to he raised the issues identified by the evaluation mission as problem areas and found little evidence from the field to support the views or perceptions of the mission. The chief secretary made it clear that the banks had limited resources to spare for financing the portfolio of opportunities of SHGs. The SAPAP team, under the dynamic and inspired leadership of Raju, gave an excellent account of itself both in the field and in articulating the vision of the Indian component of the programme. Thus, achievements in such a short period were highly commended by Henning compared to what he had seen in October 1995. The people in the country office, especially Neera, were rightly proud of what the SAPAP had achieved in India, and I was happy to report to the UNDP Country Representative Hans Von Sponeck that his efforts had not been in vain and that on the eve of his departure from India, he would be satisfied.

The instruction sheets prepared by the SAPAP team for literate as well as illiterate village activists were most impressive. One of these on institutional maturity index (IMI) of community organizations was so good that the government of Andhra Pradesh had adopted it for use throughout the state. The SAPAP had it printed on attractive and lasting plastic sheets. Most of the NPCs from other South Asian countries took specimen sheets for adaptation in their respective SAPAP areas.

Next the members and activists of the Husainipuram village organization explained to the visitors what each activist was doing in the fields of health, livestock, agriculture, water access, and so on. Initially they gave the impression of doing these services out of the goodness of their hearts, but some probing established the principle that for sustainability the activists would have to be remunerated for services provided. In fact, some of them admitted that they were already being remunerated. The visit to the MV Foundation in Ranga Reddy district to see the release of bonded child labour being assisted by UNDP was most heartwarming. The slogan of the foundation that "every child out of school is doing child labour – bonded or otherwise," has encouraged 50,000 children over the last ten years to be in school. The parent-teacher associations have played a remarkable role in attaining this objective. It was heartening to note one of the children asking the collector of the district, a lady, as to why she was not enforcing the anti-child labour laws in her district. The collector promised to do so. The secretary health of the state government also seemed to be taking an intense personal interest in tackling the problem.

Minister of rural development Mr. Jairam Ramesh echoed the views formed by Henning when he wrote,

Shoaib Sultan Khan pioneered the concept through his landmark Aga Khan Rural Support Programme (AKRSP) in Pakistan. With the support of UNDP, he introduced the concept of poverty reduction through community organizations through the South Asia Poverty Alleviation Programme (SAPAP) in India and other South Asian Countries. The Pilot initiative of SAPAP led to a full-fledged programme based on the principles of development through community organization under the Society for Elimination of Rural Poverty (SERP) in Andhra Pradesh with the support of World Bank. The project has seen the journey of over ten million organized households from despair to hope, from diffidence to confidence, from subjugation to empowerment in the last decade. The spirit of community empowerment has continued to thrive in India since then. At the behest of Rahul Gandhi (parliamentarian from Amethi), he made a visit to the Rajiv Gandhi Mahila Vikas Pariyojana (RGMVP) villages in Uttar Pradesh and helped the project to reformulate its vision. RGMVP founded on the same set of core values and beliefs and thus has much to benefit from Shoaib Khan's vision and efforts.[5]

Addressing 17th Annual Session of Sustainable Development at UN General Assembly (2009)

Figure 4.3 Addressing 17th Annual Session of Sustainable Development at UN General Assembly (2009)

Continuity of leadership: the key ingredient for scaling up the programme from Andhra to the Indian Union

The Indian component of the RAS/96/600 project has had a chequered history. After my initial euphoria about the success of the programme in Andhra Pradesh because of the willingness of the bureaucracy to act like an NGO and the availability of considerable resources to help the poor in the shape of umpteen poverty alleviation programmes and the presence of a large network of NGOs that were willing to forge a partnership with the project to reach the poor, there was a period of gloom. The programme seemed to be doomed, but a change in the stewardship of the programme brought about a remarkable transformation in the direction and vision of the programme. When a new NPC was being selected, the only request I made to Dr. Rajaji, the chief secretary of Andhra Pradesh, was to refrain from changing the NPC, as the RAS/96/600 project would not be able to survive it. True to his promise, Dr. Rajaji, helped by Venugopal, chose a person for the post of NPC whom I found completely in consonance with the vision of the project. Mr. K. Raju, during my village visits with him, exhibited perception, understanding, and commitment to the poverty alleviation strategy which left me immensely impressed and fully confident of the future of the RAS/96/600 project in India. UNDP cannot but be extremely indebted to Dr. Rajaji for the keen interest and invaluable support he provided to the programme. The contribution of Mr. K. R. Venugopal in bringing about the change in the stewardship of the programme was immense. The planning secretary, Dr. Sarma, had painstakingly tried to steer the programme out of pitfalls and continued to do so.

Neera brought all her intellectual capabilities to bear in making a strong case for additional resources from every quarter, country office, and government and regional bureau. She and her staff were a real asset to the RAS/96/600 project. In 2011 the government of India allocated US$5.1 billion for National Rural Livelihoods Mission, one of the world's largest initiatives to improve the livelihoods of the rural poor and boost the rural economy. It aimed to make a multidimensional impact on the lives of India's rural poor by mobilizing them, particularly the women, into robust grass-roots institutions of their own where, with the strength of the group behind them, they will be able to exert their voice and accountability over providers of educational, health, nutritional, and financial services. This, based on past experience, is expected to have a transformational social and economic impact, supporting India's efforts to achieve the Sustainable Development Goals on nutrition, gender, and poverty.

Notes

1 This paper is based on my notes for record of visits to India from January 20–27, 1996; June 25–July 8, 1996; December 26, 1996–January 2, 1997; June 10–17, 1997; November 5–10, 1997; April 30–May 10, 1998; and May 5, 1998; and World Bank Independent Evaluation Group. (2015) *The Million Women and Counting: An Assessment of World Bank Support for Rural Livelihood Development in Andhra Pradesh, India.* Washington, DC, World Bank.
2 Mandal is the lowest tier of local government in India.
3 NFR February 5, 1996; February 13–17, 1996.
4 NFR June 18, 1997.
5 S. K. Khan. (2015) *Speech at EU success.* Rural Support Programmes Network (RSPN), 25 November. Available at: http://www.rspn.org/wp-content/uploads/2015/12/Chairman-RSPNs-Speech-at-EU-SUCCESS-Launch-Karachi-Nov-27-2015.pdf

Reference

World Bank Independent Evaluation Group. (2015) *The Million Women and Counting: An Assessment of World Bank Support for Rural Livelihood Development in Andhra Pradesh, India.* Washington, DC, World Bank.

5   An inspirational story

Galvanizing local action for realizing global development goals: the story of BRAC making the 21st-century drive to eradicate extreme poverty at local levels

Nipa Banerjee

Bangladesh, a fairly young nation, stands steady as a confident, visionary, and forward-looking nation. The development trajectory of this low-income country, on its way to graduation to a middle-income country, is globally recognized as a unique success story, in reaching most of the global development targets (MDGs) and enthusiastically embracing the post-2015 SDGs. A Bangladeshi development NGO, BRAC is one of the driving forces behind this progress. A young visionary, popularly known as Abed, in the early years of the 1970s, founded BRAC, with the mission to empower people and communities in situations of poverty, illiteracy, disease, and social injustice. BRAC's interventions have contributed to large-scale, positive changes through the implementation of cost-effective community-level economic and social programmes that enable men and women to realize their potentials and use their assets to come out of poverty. The story of BRAC's work in development delivery highlights the benefits to be derived from working in partnership with the poor and the disenfranchised and responding to local needs for sustained impact on poverty reduction.

Keywords

Bangladesh, BRAC, community development, economic development, microcredit, conscientization, Sustainable Development Goals

Birth of BRAC

Bangladesh, with a miraculous record in enhancing the quality of life of the poor, serves as a role model for developing nations. The miracle worker most influential in defining Bangladesh's pathway to development and pulling people out of poverty is Fazle Hasan Abed.

Bangladesh, unkindly called a basket case as it emerged from a bloody war of independence in 1971, today presents not only an unexpected success story in development but also showcases a rare instance of a post-conflict country's exit out of the fragility trap. At war's end in 1971, from the devastations and destruction emerged a unique community of non-profit, non-governmental development organizations. They adopted the tool of social mobilization, which helped pave the path to development for the new nation by organizing the poor and empowering them to lift themselves out of poverty and powerlessness.

Paulo Freire's notion of conscientization of the poor is at the base of the social mobilization that guided the non-governmental organizations (NGOs) in Bangladesh (Ali, 2013). Conscientization for the empowerment of the poor involves a process of development of critical consciousness in poor communities about the social realities, the causes of powerlessness and poverty. The consciousness thus raised, motivates the communities to plan and take the required actions to help themselves attain freedom from want and poverty.

The lead subscriber to the value of the empowerment of the poor in Bangladesh, through consciousness-raising, is BRAC. Strangely enough, BRAC was founded by a 34-year-old rich and powerful young executive, employed in the Shell Oil company, Fazle Hasan Abed. In 1970, a cyclone, the most devastating ever recorded, hit this young man's homeland of East Pakistan (now Bangladesh, then a province of Pakistan), leaving its trail of over 300,000 dead and several hundred thousand more homeless. Fazle Abed, a young English-educated cost management accountant, leading a comfortable life, was shaken by the devastations he witnessed. He was shocked to see human bodies floating in the shallow waters of Bay of Bengal. He formed a volunteer group to help people rebuild (Aizenman, 2019).

Further death and destruction followed the trail of the cyclone through the year 1971, as Pakistan, in a negative response to East Pakistan's demand for self-determination through a legitimate democratic process, launched a ruthless attack on the poverty-stricken people of East Pakistan. After a bloody war spanning over nine months, Pakistan surrendered to the liberation fighters of East Pakistan and the new nation of Bangladesh was born.

Death, destruction, and suffering hit millions of men, women, and children and close to 10 million refugees returned home with little hope but to live in destitution prompted Fazle Abed to resign from his position with Shell Oil. He sold a property he owned in England and invested the resources in meeting the urgent need of rehabilitation and recovery of his nation. Thus, he founded, in 1972, the non-governmental organization (NGO) BRAC, which at the time of its birth, stood for Bangladesh Rehabilitation Assistance Committee (Zillman, 2016). The original objective of the organization, as the name implies, was to provide relief and reconstruction support. Materials and finances for house construction were provided to the refugees returning from India to a remote, almost inaccessible village, Sulla, on the north-eastern border of Bangladesh.

From rehabilitation to development

While BRAC workers were pleased to see people settling in reconstructed homes, they were concerned that a longer-term livelihood was not secured for them. Relief provided short-term support only. Inspired by the spirit of addressing the poor's longer-term needs, BRAC launched, in 1973, an integrated community development programme encompassing 200 villages to help with longer-term development. Thus started BRAC's journey in search of development, much before the dialogue on global development goals took shape.

To mark the advance from rehabilitation to longer-term development, BRAC was renamed Bangladesh Rural Advancement Committee. The aim was to foster sustainable livelihood practices in areas such as agriculture, horticulture, and fisheries. To enable the communities to lift themselves out of poverty, an integrated package was to be provided, offering services in healthcare, family planning, adult education, and vocational training (Afsana, 2012). BRAC's evaluation of its own work, however, showed that satisfactory development did not follow as the communities remained somewhat passive, not taking adequate advantage of the opportunities provided by BRAC for sustainable livelihood building. The need for stronger and upfront efforts to promote active participation of villagers in determining their own future was recognized by BRAC workers.

The lesson learned inspired young Abed to prompt BRAC to closely embrace Paulo Freire's conscientization process of agency-building through a participatory approach. Such a process would enable the poor and the socially disenfranchised to organize themselves for gaining the power to change their own lives. Young Abed, inspired by Freire's ideology, believed that communities that invest in understanding their own life situations through intra-community dialogues are able to make informed decisions and collectively plan and implement activities to address the challenges they encounter. Such communities recover fast from impacts of conflict and disasters, and are able escape from intense poverty traps. He dedicated BRAC to mobilize the poor communities of Bangladesh, with the aim to raise their abilities to lift themselves out of poverty and powerlessness.

Social mobilization promoted by a vibrant NGO community in Bangladesh, led by BRAC, has been the most effective enabler of the young nation's journey of recovery from the devastations of the war of 1971. BRAC's social mobilization approach and low-cost public awareness–building programmes about the availability and need for utilization of the existing government services has facilitated the expansion of government development initiatives, as well, and ensured nationwide outreach of these programmes.

For addressing the dual objective of poverty reduction and empowerment of the poor, BRAC follows a holistic development approach, combining economic as well as social and human development support. Employment and income generation, provision of capital, loans and market access – these are all important parts of economic development. Employment and income generation support is provided through training the poor in employable skills, getting them ready for the job market. More targeted training is provided to help the poor to earn from investing in small income-generating ventures. The economic development component is combined with human development support, which provides essential healthcare and children's education, as well as other social development programmes, with a focus on women's empowerment (Banu et al., 2001).

Organizing rural poor communities

With experimentation, BRAC learned that comprehensive multidimensional development could be best planned and delivered through organizations of the poor. Thus, BRAC adopted an approach of mobilizing the poor in forming village organizations (VOs) under the aegis of BRAC. The VOs are in the lead in planning, initiating, and managing group activities for economic, human, and social development.

The philosophical underpinning of the VOs' development operation nurtured by BRAC is that liberation from powerlessness and want must come from within the communities. The VOs provide the platform for dialogues in the communities on community issues and needs, the possible solutions of the issues, and the means required to address these. Each VO has a male and a female group. All VO members have access to BRAC's rural development programme core package of economic and social development and can be engaged, on a selective basis, in productive activities in a number of sectors, such as agriculture, social forestry, fisheries, livestock, sericulture, and so on, for income-generation purposes.

Multidimensional programme components

BRAC's economic development programme provides the foundation for all of BRAC's development support to the VOs.

Microcredit, a component of economic development, was started in 1974 to provide financial resources to meet individual basic needs or to invest in small income-generating activities. The programme expanded rapidly to millions of people with a consistently high payback rate of over 95%. The microcredit is used by BRAC as a tool especially to get the women free from the sociocultural, technological, and structural constraints they encounter, and to promote their economic empowerment by engaging them in income-generating activities. The income provides the women a position of dignity in the family and the society as they not only contribute to household incomes but also add to national productivity.

BRAC soon learned that the absence of markets can constrain microcredit borrowers' ability to use the loans productively. BRAC, therefore, started educating its loan recipients about how to invest productively in small enterprises and also began to link these enterprises with consumer markets.

BRAC invested in commercial enterprises such as the retail handicraft chain store Aarong, which links poor rural artisans with expanding urban markets. BRAC dairy and food is another such enterprise that has an established chain of milk procurement from VOs and sells processed milk and dairy products to urban customers. Rural creditors, who purchase and raise cows for milk, gain from an expanded market offered by the BRAC dairy enterprise. Poultry farms and feed mills supported by BRAC generate employment and income and facilitate training and technical support of rural entrepreneurs' microenterprise activities. So do the BRAC-supported sericulture, agro-forestry, fisheries, and livestock programmes.

These enterprises are social enterprises. The profits earned make the enterprises financially sustainable with parts of the surpluses reinvested in BRAC's development programmes, accelerating social and economic uplifting of the poor. The social enterprise programmes, thus, ensure financial sustainability of both BRAC as an organization and BRAC's social and economic programmes, and further, reduce dependence on external funding. Today, BRAC is 85% self-reliant.

Institution- and enterprise-based apprenticeship and training are provided to microfinance recipients for productive use and management of the loans and the microenterprises. Training in small enterprise development and technical and vocational training in employable skills are offered to all VO members for income and employment generation. Training provides opportunities for members of poor families to find employment abroad, as well. Bangladesh sent over 1 million skilled workers abroad in 2017. The remittances sent by those working abroad to their families help reduce rural household poverty and contribute to the gross national income of Bangladesh.

BRAC's health programme actions, predating the millennium development health goals, have always been the largest component of BRAC's multifaceted poverty reduction programme. These health activities are now the local faces of past global millennium and current sustainable development health goals and targets and thus deserve special attention.

Since 1972, BRAC has been providing preventive, curative, and rehabilitative services; BRAC's approach to providing health services is based on the simple logic that to earn an income people must be healthy. Healthy communities are, thus, fostered through bringing health to the doors of the VOs and the village households. A trained cadre of volunteer health workers (mostly women) from the villages they serve make regular visits to the VO offices and community households. Health facility–based services are also provided to address special needs.

Today, BRAC delivers comprehensive healthcare to millions. The health programme integrates awareness-building around health and nutrition, family planning, and contraception issues. The programme also delivers simple healthcare, addressing common ailments. The volunteer health workers sell essential health items such as iodized salt, dehydration salts, antibacterial soap, and contraceptives, items useful for families but regularly unavailable in rural areas. This cadre of frontline health volunteers forms the backbone of delivery of health services from the grass-roots to the national level, as they collaborate with the government, as well, in the delivery of public health services. The health workers take ownership of programme delivery to their clients, making the programme sustainable. Income generation from the sale of essential medical items serves as an incentive for the volunteer workers (Ahmed, 2008).

The BRAC health programme earned national acclaim with its nationwide implementation of Oral Therapy Extension Programme through training women in the use of Oral Rehydration Therapy (ORT). Nationwide rolling out of ORT, a low-cost means of rehydration in cases of diarrhoea, which used to kill hundreds, especially children, was one of the main contributors to the drop in infant and child mortality in Bangladesh. Another was BRAC's expansion of immunizing children against diseases.

BRAC's health volunteers' contributions to the National Tuberculosis Control Program (NTP), supported by the Geneva-based Global Fund to Fight AIDS, Tuberculosis and Malaria and implemented in collaboration with the Bangladeshi government and NGOs, have also been highly acclaimed. TB detection requires close monitoring, which could only be done through regular visits to VO households. BRAC's core health volunteers' pioneering fieldwork promoted TB detection at early stages and provided for treatment and the necessary follow-up. The patients are charged a small initial fee that would be partly returned to them after successful treatment. This element of returned fees serves as an incentive for the patients to carefully observe the instructions and the follow-up schedule established by the health workers.

BRAC pays special attention to educating children, creating the foundation for income generation and human development for nation-building purposes. BRAC began its journey with building 22 one-room schools for non-formal primary education for the poor rural areas which had no education facilities. The programme has expanded, today providing primary education to 3.8 million children, over half of them girls. BRAC school enrolment and retention rates are higher than those in government schools. The special features of BRAC's education programme include a focus on girls' education; training and recruitment of teachers from the village where the school is based to ensure retention of the teachers; timing of classes in consideration of the need for children to help in the household; and free schooling for the poor but charging fees for children from wealthier families as a form of subsidization the education of the poor.

Outside its own children's education programme, BRAC is attentive to the need for sustaining the literacy level of the population and has established thousands of libraries and reading clubs in villages across the nation. Introduction of computers and information technology has been facilitated as well.

Gender equality principles are integrated into BRAC's multidimensional programme approach. Girls and women are priority clients for all of BRAC's economic, social, and human development programmes. Special programme arrangements are made for ultra-poor women. A unique component of the BRAC programme, generating unprecedented results for women, is raising awareness of women of their rights, as enshrined in the country's constitution – reproductive health rights and rights of participation in the country's social, economic, and political processes. Women are made aware of legal services that can be accessed to protect their rights. This is a component of the conscientization process made especially for women.

Three programmes introduced in the new millennium are the Ultra-Poor Program for the elimination of extreme poverty, Pro-Poor Urban Development Program, and Climate Change and Emergency Response. Nearly 13% of Bangladeshis live in extreme poverty, earning less than US$1.90 a day. The ultra-poor are the lowest-earning subset of this population, and they lack access to basic social safety nets and hardly receive any government and NGO services. In 2002, BRAC pioneered aiming the Ultra-Poor Program at addressing the needs of this group of severely marginalized poor. The intervention, named the Graduation Program (from ultra-poor status), assists the community of extreme poor through a blended programme of livelihood promotion, social protection, and financial and social inclusion. BRAC introduced a tailored approach to reach this group. The delivery package to the target group includes a modest stipend to the beneficiaries; grants and asset transfers for basic needs satisfaction; and interest-free loans along with training for micro-investments with potentials of income generation. Training is provided in employable skill development, as well.

BRAC's health programme especially addresses the needs of the ultra-poor and mainly women in this group. The ultra-poor are the most vulnerable and susceptible to diseases as they lack the basic necessities of life such as food and shelter, and they suffer from extreme malnutrition. BRAC introduced a tailored approach to reach this group, providing skill development training to promote income generation and basic healthcare. A modest stipend is provided to the beneficiaries. Basic healthcare includes essential immunizations; nutrition, sanitation, and family planning education; and a supply of safe water and essential drugs. Women are provided with prenatal and postnatal care. TB control and treatment of basic ailments are provided by specially trained health providers, who are equipped with supplies of basic drugs and equipment. Referral services have been established to expand the ultra-poor's access to special services that the government and other agencies provide.

This multidimensional programme, tailored for the ultra-poor, aims to foster self-reliance and to help families transit out of extreme poverty and join BRAC's mainstream programme of poverty reduction. The Ultra-Poor Graduation Program is now a centrepiece of the BRAC programme. To date the programme has served 1.9 million ultra-poor households in Bangladesh.

Rapid urbanization presents challenges as it is known to swell the numbers of urban poor. Approximately 20 million people now live in urban informal settlements in Bangladesh. It is estimated that by 2050, 50% of the total population will be urban dwellers. BRAC's work with people living in urban poverty involves partnership with the government, non-governmental organizations, and private sector with the goal of improving urban life and conditions, which will have a beneficial impact on the lives of the urban poor. BRAC's work comprises direct delivery of basic services to the poor and connecting them with government and non-governmental service agencies for further coverage. In parallel, a conscientization process builds awareness in these communities of their rights and entitlements and develops their capacities to demand implementation of pro-poor policies and improved service delivery. Support is also provided to develop capacities that would help the poor to benefit from urban growth.

Climate change is an issue that severely affects Bangladeshi people and the country's economy. Following its holistic approach to development, BRAC works to strengthen the resilience of vulnerable communities against the impact of climate change and the disasters it brings. Mitigation activities that combat adverse impacts of climate change on agriculture and small farmers are planned and implemented. The programme also includes measures for disaster management and organizing humanitarian responses to emergencies.

Better spending principles

Better spending features of BRAC programme are several:

- Pursuit of Paulo Freire's ideology of conscientization of the poor that generates a consciousness of the social, economic, and political processes that impact on community life and empower the communities to discuss and decide the measures that they can take collectively to improve their own lives and sustain the gains. Connected in spirit with this ideology is BRAC's village-based social mobilization process, empowering poor communities to determine their own destinies;
- BRAC's approach to learning as it innovates. BRAC values lessons learned and using these in redefining programme strategies;
- BRAC's holistic approach to poverty reduction, through economic, social, and human development: microcredit for income generation; training for employment and income generation; provision of health services; children's education; and gender equality;
- BRAC's priority focus on the ultra-poor with the aim to eliminate ultra-poverty by 2030;
- BRAC's primary focus on women, promoting their participation in the development process through awareness-building and empowering them with education, income, and better health. Eighty-seven percent of BRAC clients are women;
- BRAC interventions to promote two-pronged sustainability: sustainability of the impact of BRAC programmes on the beneficiaries and sustainability of BRAC through implementation of social entrepreneurial programmes that will fetch income for the beneficiaries and BRAC;
- BRAC's philosophy of starting small and scaling up to reach as many poor people as possible (Ahmed & French, 2006). In every area of BRAC's development intervention and social mobilization process, efforts are made to replicate and scale up small-scale benefits for greater impact nationwide. The number of beneficiaries reached show the scaling-up effect; and
- Conscious focus on local agenda for development, drawing inspiration from the global agenda – the Millennium and Sustainable Development Goals combined with reflections on the needs and means to contextualize the goals and the accompanying activities to suit country-specific situations.

Outreaching global development goals

A cursory glance at some of BRAC's achievements in 2017 only leaves no doubt that the BRAC-adopted better spending tools are generating results, with a steady aim to reach the SDG targets. Today, Bangladesh's position in the MDG track presents a happy picture, an amazing success story of a 47-year-old nation that started its journey from steep poverty, with few assets and caught in the fragility trap. BRAC's common-sense but innovative programmes have helped move Bangladesh to attain the majority of the MDG targets, a rare feat. Fazle Hasan Abed, the man who founded BRAC when he was 34 years old and lead it for close to 50 years without losing his innovative spirit and dedication to the poor, remained committed to focusing on the targets of SDGs for social and economic transformation of the poor and the powerless.

Addressing the MDGs and reflecting the spirits of many of the SDGs, BRAC's investment in a multidimensional development programme has reached and benefitted hundreds and thousands of the poor and vulnerable, especially women and children, as is clearly reflected in the statistics collected for the year 2017 alone.

The sheer number of women reached shows astonishing results in gender-focused programmes for promotion of gender equality, a global goal under MDGs, as specified in SDG 5. Eighty-seven percent of the microfinance clients are women. Close to 25,000 female clients accessed basic legal aid services. Approximately, ninety thousand women graduated from human rights and legal education courses. Over 19,000 survivors of domestic violence, most of whom were women, received emergency medical support. Over a million people, of whom 53% were women, were sensitized to the cause of ending violence against women and children. Close to 49,000 leaders from women-led grass-roots institutions participated in the local organizational structures. Nearly 79,000 secondary school students' awareness was raised about the impact of child marriage. Ultra-poor women receive special attention.

Outreach of the health programme and contributions to health-related MDGs and SDGs are no less impressive. There is an emphasis on fostering good health and well-being (SDG 3); 90% of poor households received basic healthcare services. A million children received micronutrient packages. Fourteen million women and married adolescents were counselled on family planning. Nearly 15 million adolescent girls were counselled on menstrual hygiene management. Over 3 million women and adolescent girls received nutrition education. Attending to the MDG goal of eliminating life-threatening diseases, over a million TB presumptive cases were tested and over 100 diagnosed. Ninety-five percent of TB cases were treated successfully. Over 19,000 malaria patients were treated. Thousands accessed safe drinking water, and hygienic toilets contributing to SDG 6 (provision of clean water and sanitation); hundreds of schools received improved access to latrines and drinking water sources.

BRAC activities have assisted Bangladesh in reaching the MDG education goal target, establishing village schools and expanding primary education. An exemplary number, close to 4 million children, over half of them girls, were enrolled in schools in 2017. Nearly 100% of students completing the primary school curriculum graduated. Over 60,000 students from hard-to-reach households completed primary education through BRAC's non-formal primary education programme. Millions of children across Bangladesh were engaged in reading, socialising, and activity-based learning in BRAC's adolescent development clubs, travelling libraries, and multipurpose community learning centres. Thousands of children with special needs enrolled in BRAC schools. Close to 400,000 children between the ages of 3–5 years accessed early childhood development and pre-primary programmes.

BRAC's programme made a substantial contribution to Bangladesh attaining the MDG target in poverty reduction. It is currently providing comprehensive support to SDG 1 and 2 (respectively, no poverty and zero hunger) and also goal 8 (decent work and economic growth). The microcredit programme reached over

5.5 million people in 2017 alone, an increase of 6% from 2016. Contributions to employment generation, and thus income, are not negligible, with close to 34,000 people having received skills training for employment.

Further, BRAC's programme in support of the 2030 Agenda of total elimination of extreme poverty has gathered enormous momentum. One dollar of investment in the targeted ultra-poor programme is bringing a 5.4 time return in income and assets over a period of seven years. Seventy-five to ninety eight percent of the ultra-poor participants graduate from the programme within 18 to 36 months. Randomized control trials show that income and well-being of the targeted ultra-poor group is better than that of those who were not members of the programme. Those served by the programme had more income and more to eat, and enjoyed better mental health. The ultra-poor groups had 37% higher earnings than those in the control group. Over 75,500 of these households graduated, bringing the total graduating household number close to 2 million. The UN secretary general advocates for all developing countries to adopt a graduation programme to meet the poverty reduction target set under the SDGs.

In support of multidimensional poverty reduction in Bangladesh's hard-to-reach areas, 90,000 households received assistance for recovering from devastating floods. Ninety percent of the households received health services and 50% had access to financial services in these areas. Over 99% of the primary school children from such areas graduated.

BRAC works to attain several of the targets set by Sustainable Development Goal 13 (climate action), 14 (life below water), and 15 (life on land). BRAC actions reduce risks of increased poverty incidents resulting from environmental degradation and also protect the vulnerable from falling back into the intense poverty trap. BRAC follows a two-pronged programme of humanitarian and relief responses for the vulnerable adversely affected by disasters caused by climate change. BRAC also has a programme to combat the impact of climate change on agriculture, food production, and food security.

Close to 200,000 families, with a large percentage of women included in the group, were reached through BRAC's holistic humanitarian response to crisis resulting from climate change. Hundreds of thousands of small farmers threatened with lowering staple crops yields, resulting from climate change risks, accessed BRAC's agricultural services and were trained in high-yield and climate adaptive crop production technologies. They adopted model farming practices to overcome climatic risks.

BRAC's pro-poor urban development programme contributes to poverty reduction related to SDG 11 (sustainable cities and communities). Over 70,000 dwellers received a wide range of development support. Thousands received livelihood support and education grants. Hundreds of street children in Dhaka received services at drop-in centres. To meet the minimum housing need, over 5,000 houses were rebuilt. Relief support was coordinated in partnership with local community organizations, municipalities, and the private sector. This approach connects the BRAC programme to Sustainable Development Goal 17: building partnerships and coalitions for development financing. BRAC, therefore, is positively responding to the need for supporting a new development financing landscape, essential for addressing the ambitious 2030 Agenda (Kashem, 2015).

BRAC stands on guard delivering Sustainable Development Goals in Bangladesh

Bangladesh, immersed in steep poverty over several decades, has boldly embraced the global goals of development set in the beginning of the new millennium. It stands today as a development model, having completed the MDG track in 2015, with more than satisfactory results in the attainment of the majority of the global goals at the national level. These include increased access to education and health: increased immunization coverage and disease control; gender equality; declining maternal and infant mortality; heightened life expectancy; and food security.

The country's success in localizing global goals is very much entwined with the efforts of the local leader, Fazle Hasan Abed, who, embracing Paulo Freire's philosophy of conscientization of the poor, inspired the disadvantaged and the disenfranchised in one of the poorest countries of the world to organize themselves in fighting poverty and powerlessness, which are at the centre of global development goals. The members of these rural poor communities, the unknown soldiers of the early fight against poverty and inequalities, should not be forgotten. So we should also remember the story of BRAC, a very small, aid-dependent organization's transformation into the world's largest NGO; it is 85% financially self-reliant, with an annual budget of 2 billion dollars, and remains dedicated to the cause of fighting poverty.

As laid out in BRAC's five-year (2016–2021) development strategy, the organization's current development agenda at the national level is guided by the global agenda 2030 and the priorities of the Bangladesh government's seventh five-year plan (BRAC, 2018). The latter provides the contextual base for planning and acting on realizing the SDGs at the national level.

The principles of better spending will be at the base of all BRAC actions. Of the 17 SDGs, BRAC's focus is on those that directly impact extreme poverty eradication in all its dimensions, with an emphasis on improving the quality of basic services and of education; these have been identified as the priority challenges encountered by Bangladesh.

The contextual analysis also highlights the urgent need for sustained efforts in reducing maternal mortality as data shows that some of the impressive gains of the MDG years in achieving the target in this category have stalled. BRAC undertook research, revisiting the past programs and recording lessons from the past, to understand why the earlier enormous improvements in effective handling of the maternal mortality issue had stagnated. The research finding is that the interventions for reduction of maternal mortality had been failing to reach the vast majority of women. The better spending principle of scaling up to extend outreach was missing in the program. The hard reality is that the SDG target of 70 deaths per 100,000 live births can be reached by Bangladesh by 2030 only with coordinated actions of the government and the civil society, both scaling up the outreach of the interventions. In effect, the achievement of the sustainable development goal targets in all sectors would remain a dream without local partnership-building between and joint actions by the government institutions, the private sector, the civil society, and the people (Neupane, 2017).

Here lies the story of world's largest development organization in terms of its outreach and outcomes. The organization continues to work relentlessly to translate into practice, at local levels, the global dreams of ending poverty, inequality, and injustice, ensuring healthy lives and education, and contributing to prosperity and economic well-being of the people.

Tracking BRAC results delivering global development goals (calculations based on BRAC Report 2017)

Gender equality	• 87% of microfinance clients are women • 24,599 women accessed basic legal aid services • 90,025 women graduated from human rights and legal education courses • 1,775 female survivors of violence received legal aid and counselling • 1.2 million people living in poverty (mostly women) accessed social safety net services • 19,367 survivors of domestic violence (mostly women) • 48,766 leaders from women-led grass-roots institutions participated in the local power structure • 78,684 secondary school students educated about awareness of sexual harassment and child marriage • 1.3 million people (of them, 53% were women) sensitized on ending violence against women and children • 585,906 men and boys engaged in changing gender roles and positive parenting • Developed capacity of 15,135 members in community-based forums to be agents for gender-transformative change • 2,273 women participated in local power structures • 110,267 women supported through post-disaster recovery and rehabilitation initiatives towards building resilience
Access to healthcare	• 1 million children received micronutrient packages • 3.7 million women and adolescent girls received nutrition education • 14 million women and married adolescents counselled on family planning • 2.7 million maternity services provided through BRAC outreach sessions • 169,292 appropriate reading glasses provided and 29,862 cataract surgeries performed • 43,714 people accessed safe drinking water • 207,404 people gained access to hygienic toilets • 1.49 million adolescent girls counselled on menstrual hygiene management • 110 schools received improved latrines and drinking water sources • 1.35 million TB presumptive cases tested and 162,219 people diagnosed • 337,303 malaria cases tested and 19,145 people diagnosed and treated

(Continued)

Access to education	• 3.8 million children, over half girls, enrolled in 43,973 schools and centres • Drop-out rate shows as 6% • To date, more than 11 million children have graduated from BRAC schools • 396,931 children between the ages of 3–5 years accessed early childhood development and pre-primary programmes • To date, over 6 million children have completed pre-primary schooling • 1.4 million children across Bangladesh engaged in reading, socializing, and activity-based learning in adolescent development clubs, travelling libraries, and multipurpose community learning centres • 44,898 children with special needs enrolled in BRAC schools and centres
Poverty reduction	• Every US$1 invested in the programme targeting the ultra-poor resulted in US$5.40 in income and assets over a period of seven years • 75%–98% of the participants globally meet the country-specific graduation criteria in 18–36 months • 75,658 households graduated from ultra-poverty in one year, bringing the total to 1.8 million In Bangladesh's hard-to-reach areas: • 60,000 households supported after devastating flash flood and an additional 33,000 households supported through flood-recovery initiatives • 90% of households received basic health services and 50% had access to grants and assets • 99.91% of students completed primary education
Employment and income generation (employable skills and training in microenterprise development)	• 33,980 people supported with skills training, jobs, and decent work interventions • 1,425 people trained under the skills for employment investment programme and empowering ready-made garment worker projects
Climate change	• 170,875 people reached through holistic humanitarian response and climatic interventions • BRAC adopted a climate change strategy, an environment and social safeguard framework, and an environment policy • 443,066 people accessed agricultural services • 160,000 farmers trained in higher-yield and climate-adaptive crop production technologies
Pro-poor urban development	• 71,000 people in 300 slums accessed a wide range of services • 12 community information resource centres established • 300 community development organizations, community action plans, social mapping, and well-being analyses completed to improve quality of life in slums • 1,926 people received livelihood support and 2,277 received education grants • 300 children living on Dhaka streets received services at drop-in centres • 5,500 houses were rebuilt, and relief coordination and support were provided in slums in partnership with local community organizations and the City Corporation

Source: BRAC's 2017 Annual Report

References

Afsana, K. (2012) Empowering the community: BRAC's approach in Bangladesh, in Hussein, J., McCaw-Binns, A. M., & Webbe, R. (eds.), *Maternal and Perinatal Health in Developing Countries*. Oxfordsire, CABI, pp. 170–180.

Ahmed, S. M. (2008) Taking healthcare where the community is: The story of the Shasthya Sebikas of BRAC in Bangladesh, *BRAC University Journal*, V(1), pp. 39–45.

Ahmed, S. M., & French, M. (2006) Scaling up: The BRAC experience, *BRAC University Journal*, III(2), pp. 35–40.

Aizenman, N. (2019) Meet the most influential poverty fighter you've never heard of, 7 April. Available at: //www.kcur.org/post/meet-most-influential-poverty-fighter-youve-never-heard#stream/0 [Accessed 12 April 2019].

Ali, S. (2013)The dialectic of freedom, *BRAC Blog*, 6 February. Available at: http://blog.brac.net/the-dialectic-of-freedom/ [Accessed April 2019].

Banu, D., Farashuddin, F., Hossain, A., & Akter, S. (2001) Empowering women in rural Bangladesh: Impact of Bangladesh rural advancement committee's (BRAC's) programme, *Journal of International Women's Studies*, 2(3), pp. 30–53.

BRAC. (2017) *2017 Annual Report*. Dhaka, Bangladesh, BRAC.

BRAC. (2018) Implementation of 7th five-year plan and SDGs, 22 January. Available at: www.brac.net/latest-news/item/1127-implementation-of-7th-five-year-plan-and-sdgs [Accessed 10 April 2019].

Kashem, A. (2015) Development: Whose fault is it anyway, *BRAC Blog*, 22 November. Available at: http://blog.brac.net/development-whose-fault-is-it-anyway/ [Accessed 3 April 2019].

Neupane, S. (2017, September 21) Community engagement in the SDG era: Why? *Think Tank Initative Blog*, 21 September. Available at: www.thinktankinitiative.org/blog/community-engagement-sdg-era-why [Accessed 2 April 2019].

Zillman, C. (2016) 3 keys to starting a successful NGO from the world's most effective poverty fighter, *Fortune*, 2 December. Available at: http://fortune.com/2016/12/02/brac-fazle-hasan-abed-ngo [Accessed 11 April 2019].

Further readings

Ahmed, S. M., & Rafi, M. (1999) NGOs and evaluation: THE BRAC experience. *World Bank Conference on Evaluation and Poverty Reduction*. Washington, DC, BRAC, pp. 1–17.

BRAC. (2015) Sustainable development goals: Let's make it about the people, 24 September. Available at: https://medium.com/creating-opportunity/sustainable-development-goals-let-s-make-it-about-the-people-d072956999c6 [Accessed 12 April 2019].

Cho, S., & Sultana, R. (2015) Journey from NGO to sustainable social enterprise: Acceleratory organizational factors of BRAC, *Asian Social Work and Policy Review*, 9(3), pp. 293–306.

Chowdhury, A. M., & Bhuiya, A. (2004). The wider impacts of BRAC poverty alleviation programme in Bangladesh, *Journal of International Development*, 16(3), pp. 369–386.

Matin, I. (2002). *Targeted Development Programmes for the Extreme Poor: Experiences from BRAC Experiments*. Dhaka, Research and Evaluation Division, Bangladesh Rural Advancement Committee (BRAC).

Neupane, S. (2017) Community engagement in the SDG era and the role of development donors, *Think Tank Initative Blog*, 19 October. Available at: www.thinktankinitiative.org/blog/community-engagement-sdg-era-and-role-development-donors [Accessed 2 April 2019].

Ordonez, A. (2018) The time is ripe for southern perspectives, *Think Tank Initative Blog*, 7 December. Available at: https://onthinktanks.org/articles/the-time-is-ripe-for-southern-perspectives/ [Accessed 2 April 2019].

Seelos, C., & Mair, J. (2006) *BRAC – An Enabling Structure for Social and Economic Development*. Pamplona, Spain, IESE Business School - University of Navarra.

Smillie, I. (2009) *Freedom from Want: The Remarkable Success Story of BRAC, the Global Grassroots Organization That's Winning the Fight Against Poverty*. Sterling, VA, Kumarian Press.

Srivastava, L. (2013) *BRAC: A Pioneering Bangladesh Human Service Organization (1972–2009)*. Berkeley, University of California, School of Social Welfare.

UN Habitat. (2007) *People's Process in Post-Disaster and Post-Conflict Recovery and Reconstruction*. Regional Office for Asia and the Pacific, Fukuoka, Japan, United Nations Human Settlements Programme.

United Nations. (2019) About the sustainable development goals. Available at: www.un.org/sustainabledevelopment/sustainable-development-goals/ [Accessed 10 April 2019].

United Nations Development Programme. (2019) Millennium development goals. Available at: www.undp.org/content/undp/en/home/sdgoverview/mdg_goals.html [Accessed 21 March 2019].

World Health Organization. (2003) *Social Mobilization for Health Promotion*. Manila, WHO Regional Office for the Western Pacific.

6 Water management is water measurement

Fayyaz Baqir

There are thousands of abandoned water supply schemes in Pakistan due to poor operation and maintenance of these schemes by local water and sanitation departments. These schemes were built with multimillion-dollar loans adding heavy burden to the low-income taxpayers, who bear the brunt of the heavy indirect tax-based revenue system. Many government experts ask for new schemes in view of the existing dysfunctional schemes. Anjuman Samaji Behbood (ASB) has restored one such scheme through a process of connecting the dots in Bhalwal town in Pakistan. ASB mobilized beneficiary communities in Bhalwal to install meters for paying water charges and engaged and activated the government to carry out repair and maintenance. ASB restored the scheme without asking for a single penny in grant money from the government or donors. It is an innovative case of stretching the dollar for improving local SDGs.

Keywords

Better spending, water governance, sustainable development, poverty alleviation, gender and development, inclusive governance

Context: a counterintuitive approach to access water for the poor

According to the Pakistan Council for Research in Water Resources (PCRWR), there are hundreds of dysfunctional water supply schemes worth hundreds of millions of dollars, causing unbearable miseries to women who are the sole water carriers in rural households. Thousands of young girls cannot attend school because they have to fetch water for their families. The majority of low-income people, especially women, have hepatitis and other waterborne diseases due to water contamination and poor sanitation. These schemes can be restored with the provision of meager resources and low-key technical intervention. The enigma of large-scale existence of dysfunctional schemes in the presence of generous donor funding, elaborate community engagement frameworks, and political committment to deliver at the highest level was explained by Dr. Akhter Hameed Khan, a leading development practitioner, as symbolizing anarchy. By anarchy he meant that both the rent-resisting and rent-capturing actors cancelled out their mutual power in such a situation. This created a space where anyone could make the difference by opening the channels of communication between the isolated stakeholders. Dialogue can diminish the power of rent-seekers to embezzle public resources. This chapter describes one such story which unfolded in the small town of Bhalwal in the Punjab province of Pakistan.

It is important to note here that Anjuman Samaji Behbud's (ASB) approach not only dealt effectively with the government's and market's failure in provision of water to low-income communities, but it effectively dealt with a civil society failure in addressing this challenge as well. Civil society organizations (CSOs) entered the water and sanitation sector in Pakistan in 1980s as service providers. They provided subsidies for the improvement of services at selected locations. They soon realized that without engaging and activating the government they could not be successful on a significant scale, and at the beginning of new millennium they made a transition to the right based approach. Right based work called for the increased allocation of resources for water supply schemes. This approach has not had much success because there are hundreds of dysfunctional schemes, despite the increased allocation of public funds. Realizing the limitations of right based approach one social entrepreneur tested a social business model approach to restore a dysfunctional water supply scheme in the small town of Bhalwal. Managerial success of this restored scheme is based on water metering. It follows the principle "water management is water measurement." This approach has led to a success story which needs to be documented, analyzed, and presented as a business model for water-demand management by the social entrepreneurs. The business model in this context does not mean a model for privatization of services but an improved, efficient, and sustainable provision of public services to communities "within the existing system and the available means." This model shows an innovative way for aid effectiveness, sustainable water management, and mitigation of climate change. This approach is counter-intuitive because it has shown that levying user fees for receiving water services does not add to but reduces the financial burden on the un-serviced or under-serviced communities.

Political commitment

During the 1990s the World Bank and the Asian Development Bank (ADB) financed many water supply schemes in Pakistan. The World Bank also provided funding for launching a private company, Punjab Municipal Development Fund Company (MPDFC). This so-called private company was staffed by bureaucrats appointed by the very provincial government suffering from bureaucratic inefficiencies.

During this period the chief minister (CM) of Punjab Shahbaz Sharif was searching for solutions to the inefficient provision of basic services. In 1998 during his visit to the world-renowned Orangi Pilot Project (OPP), known for providing sanitation services on a self-help basis to a population of 1 million people in the informal settlement of Orangi, he came across Nazir Ahmad Wattoo. Sometime later the CM was put in exile by the military ruler Pervaiz Musharraf. On his return after ten years he was re-elected as CM of Punjab. He remembered Wattoo and invited him to a meeting on provision of water to neglected areas of Punjab.

When Wattoo entered the meeting hall it was full of people. The CM immediately recognized him and asked him for his views. He was subsequently invited to a few consultation meetings and finally the CM invited him for a one-on-one meeting. In this meeting, with tears in his eyes, the CM asked Wattoo, "Can we not even provide water to the poor?". Wattoo said, yes, "if you follow the principle that nothing is free. Your government should issue a Water Act to provide rational fee-based services for the provision of water".

The CM thoroughly discussed the proposal for the provision of fee-based water services with him. To draft the legislation for a Water Act the CM created a committee consisting of four provincial secretaries: Secretary PHED, Secretary Local Government, Secretary Health, and Secretary Urban Unit, and he appointed Wattoo as chair of the committee to ensure that the drafting work did not suffer from bureaucratic neglect. During this whole period the principle secretary (PS) of the CM, Tauqir Shah, played a key role in facilitating Wattoo's interaction with the provincial bureaucracy and the CM.

Experimentation with the template-based participation strategy

A process for introducing a participatory approach for the provision of water to low-income communities was under way in the departments dealing with water supply services in Punjab. In 2005, the Public Health Engineering Department (PHED) of Punjab introduced a policy of community mobilization for development, implementation, operation, and management of rural water supply schemes. As part of this policy PHED started its work by forming Water Users Committees (WUCs) in the target villages. WUCs were formed through elections in the general meeting of the villagers. In these meetings PHED explained Terms of Partnership (ToP) to the WUCs. Once WUC agreed to the ToP, PHED prepared a rough cost estimate for administrative approval. Copies of rough cost estimates were sent to the Sub Divisional Officer (SDO), Executive Engineer (Ex En) and Provincial Secretary PHED. From the secretary's desk this proposal was marked to the District Coordination Officer (DCO)[1] and Planning and Development (P&D) Unit of the provincial government and sent back to the Secretary after the review.

PHED was required to share a copy of the rough cost estimate with the WUC and subsequently the community development unit of PHED started the process of community mobilization known as the "Flying Visit." The Flying Visit was the approach used by the Community Development Unit (CDU) of PHED of the government of Punjab (GOP) to engage communities. It consisted of numerous steps and was used to ensure the community's participation in (i) identifying the need for water supply schemes; (ii) creating a local organization for mobilizing communities, financial resources, and local support to develop specific components of the water supply scheme and to take over Operation and Management (O&M) responsibilities of the scheme after completion. This process consisted of some or all of the following steps:

- Conducting social mapping to determine the economic standing of target beneficiaries.

- Holding a village assembly meeting to determine if the community needed a water supply scheme.
- If the community expressed the need, CDU explained the ToP to the community. Under the TOR, community had to (a) form a Water User's Committee (UC) or a Community Based Organization (CBO), both for males and females; (b) verify if it was ready to pay a 2% cost of the total scheme as its financial share; (c) provide the land for the water source, overhead reservoir, and route for the rising main; and (d) take responsibility for O&M of the water supply scheme after its completion.
- Subsequent to the acceptance of the ToP, taking water samples from potential water sources and determining the most appropriate location for drilling and depth of the pipe needed for accessing clean drinking water.

The CDU worked as a support organization for water users and trained community members for social mobilization, technical oversight of the contractors' work, financial management of the scheme, and record keeping. This was a way of working that was different from the traditional practices.

Three approaches to design and implement a rural water supply scheme

During the previous decades, water supply services (WSS) were identified in three different ways: (i) by the elected representatives like members of national and provincial assemblies – that is, MNAs and MPAs; (ii) by surveys of engineering staff of PHED; or (iii) by a community-based need assessment conducted by CDU in collaboration with the technical staff of PHED. Traditionally politicians preferred to select the site of a scheme, keeping in view their vote bank; PHED gave preference to schemes at sites having brackish underground water and a large population; community-based assessment focused on a community's willingness to takeover O&M after completion of the scheme to ensure sustainability. Things changed when the community-based approach was adopted by PHED. Whereas the first two approaches were based on the outsider's preferences and interest, CDU's approach had the potential of building local ownership and sustainable operation and management of the water supply services.

The Flying Visit was a powerful tool for the community's engagement in managerial functions. Under the proposed template for the project, the cycle management of the scheme WUC was entitled to receive a copy of the rough estimates; it was trained to conduct technical oversight effectively and get the scheme completed according to approved specifications. After the approval of cost estimates, PHED was tasked to develop a technical proposal and ascertain the community's capacity and willingness to undertake O&M of the water supply scheme.

After administrative approval, cost estimates were back to local staff at PHED and CDU so that they could prepare a detailed cost estimate which consisted of approximately 100 pages. A detailed cost estimate was then approved by the Sub Engineer (SE) and sent for technical approval to the head draftsman in the chief engineer's office. From there it was sent to P&D and finally approved by the secretary for tendering. If the contractor deviated from approved specifications, the community could refuse to take over the scheme. If the community was not kept informed on rough estimates and other critical points, things could go off track. This was a well thought out procedure to pre-empt pilferage of funds. However, this inclusive governance framework did not work.

The promises and perils of downward accountability

As part of the procurement process, tendering was done by the Executive Engineer (EX-EN). Work was allotted to the contractor by the chief engineer, who also grants the technical sanction (TS). At the time of allotment of tenders, contractors are required to deposit 5% of the scheme cost as a security deposit, refundable after completion and testing for one year. The contractor must perform the yield test at the location selected for drilling. This test is the second most critical step for ensuring access to a clean source of water. Due to weak community involvement, this test is not conducted at most of the places; the contractor receives payment based on a fake report. This is the fatal flaw. This happens because community organizations without the support and guidance of a support organization cannot check the fraudulent practices of the contractor. Contractors collude with corrupt insiders and split the spoils after being paid for substandard or unfinished work. The support organization under the ADB project was tasked to the CDU under the proposed inclusive governance framework. However, CDU staff is project staff. They hold temporary contract jobs and can be removed from service at the discretion of those very officers who are engaged in financial malpractices. Public funds can therefore by stolen with convenience at any stage of the project cycle. After completion of the scheme, capacity of water production is checked. Measurement book (MB) entries must be made. Without MB entries, the contractor cannot receive payment. It is supposed to be a confidential record, but usually it is carried out by the contractors themselves. This again violates the spirit of community ownership and can lead to a failed scheme if not properly followed. Under the rules, local government officials cannot be taken to account for the quality of the work carried out by the contractor and can get off the hook by blaming the contractor for a dismal performance (Iftikhar et al., 2018).

The contract bidding and work awarding process is guided by the sole criteria of awarding contracts to the lowest bidder. This throws out competent contractors due to higher bids for quality work. The track record of the contractor and compliance with project specifications and timelines are not factored in when accepting the bids and releasing payments. The hiring of a wrong contractor causes a waste of public and community resources. This calls for finding effective ways to deal with the contractor. One way of dealing with the contractors is to create awareness of the rules made by CDUs among the beneficiary communities. A survey of well-functioning water supply schemes revealed that WUCs were able to get quality work done by the contractor due to knowledge of the rules governing community-based O&M. One WUC based in the village of Vairo had very limited financial resources, and they did not want to take over a scheme which would become a permanent financial liability for them. As they monitored the quality of construction work, they realized that the contractor did substandard work in violation of the contract clauses. At this stage they came to know that if they refused to sign the MOU for taking over the water supply scheme from PHED, the contractor would come under pressure to deliver as agreed.[2] However, this approach did not help in dealing with the failed schemes handed over to communities without compliance with the due process. ASB found a way of dealing with this problem by designing a strategy for the restoration of abandoned schemes. ASB tested its business model in the town of Bhalwal.

Business model development in Bhalwal

ASB's President Nazir Ahmad Wattoo had learned through his experience to follow some simple and sound principles of community development: start small, find a local partner, engage in dialogue, understand local conditions, and ask for cooperation to restore broken links, not finances. When the chief minister Punjab asked him to help restore the dysfunctional water supply scheme in Bhalwal he grabbed the opportunity. He knew that if he could demonstrate his success on a small scale it would expand through self-selection by local champions, at a slow pace at first, but at a more rapid pace with the passage of time.

Bhalwal is one of the six Town Municipal Administrations (TMAs) of Sargodha district. It is a rural market town in the citrus farming area of Sargodha district. It has a population of over 77,000. The Tehsil municipal officer (TMO) is tasked with coordination and oversight of community-driven projects (PMDFC, 2005). The city was built on a grid pattern as a planned city in early part of the 20th century, during development of canal colonies by the British administration; it was built to serve as a small rural market town. It is situated in a canal-irrigated area inhabited mostly by progressive farmers. Groundwater in the town is brackish and has been contaminated due to seepage of waste disposed by the sugar industry and leeching of heavy pesticide and fertilizers from citrus farms. The first water supply scheme in the town was built in 1927, but it was inadequate to meet the needs of the growing population and became dysfunctional. Only a small portion of the population located on the canal side had access to fresh water. Later on, a raw water treatment plant also provided drinking water to the residents (PMDFC, 2005). The city was divided into various zones. By 2013 water supply was available to old city zone-1 residents for 3.5 hours per day, and to zone-2 new city residents for 2 hours per day. This was caused due to low production of water.

Water chlorination was done for overhead tanks of zone-1 only. Heavy contamination of water was reported for the entire system. Most of the consumer connections were poorly installed, passing through service drains and crossing sewer lines. Old water pipes were seven to nine feet below the surface as roads and streets were raised due to earth fill over time. Other households had to travel as far as five kilometres to buy water from vendors, paying Rs. 20 per day to purchase the water, whereas household connections were charged only Rs. 360 per year by the municipal authorities. Overhead tanks had the capacity to store 100,000 gallons and could provide water 24/7 to 2,500 household, but only 270 households out of 4,000 received connections after the scheme was completed in 2013 (PMDFC, 2005; Sahi, 2015). Piped sewerage coverage was 50%. There was no solid waste management system in the city. The city had low service quality, low level of water connections, and low recovery of water user charges. It was a case of dead investment. The CM was at loss for how to fix this problem. The first task for ASB was to look into the governance framework for the drinking water supply in Bhalwal.

Governance framework

A survey report published by PCRWR stated that 9 of the 16 water supply schemes in Tehsil Bhalwal were non-functional, and 85% of the population that was supposed to be served did not receive water. The reasons for closure of schemes included breakage of infrastructure, disputes within the community on collection of operation and maintenance charges, non-payment of electricity bills, and theft of transformers and electric motors. All these schemes were constructed by PHED. No water treatment was carried out in six out of seven functional schemes, and only 67% of the population had access to safe drinking water, due to biological contamination of water at the source. Water samples collected at the consumption point revealed that 88% of the water showed biological contamination. Tehsil (town) administration was collecting water supply charges through revenue recovery in 43% of the schemes only (Tahir et al., 2011). The water delivery system was suffering from deficiencies at both the supply level and demand level. It was as if the system was trapped at a lower level of equilibrium and had the potential to perform at the higher level to the benefit of everyone. The question was how to break out of this cycle. The government of Punjab tried to solve this problem by providing funds and technical advice through a private company, Punjab Municipal Development Fund Company (PMDFC). As described below, PMDFC did not make any headway in dealing with the problem.

Provision of finances and technical assistance by PMDFC

PMDFC was established as a private company to undertake Punjab Municipal Services Improvement Programme with the financial assistance of the World Bank. The programme also aimed at building the capacity of Town Municipal Administrations (TMAs) to improve the infrastructure facilities and service delivery in small towns. A detailed grant funding system was established to allocate resources to TMAs. PMFDC's mission for assessing the existing situation in Bhalwal found out that TMA did not have the capacity to prepare and implement plans for infrastructure development and service delivery due to understaffing. TMA's documentation, data collection, and mapping work was deficient and obsolete. It did not have the knowledge and capacity to plan according to local needs. PMDFC proposed to fill the capacity gap by upgrading the skills of existing staff and hiring a host of consultants. TMA also displayed a weak financial management capacity. PMFDC approved grant funding for the water supply scheme in Bhalwal after detailed technical preparation. Documents submitted for a grant application included an inception report, a set of guidelines on methodologies and approaches to data collection, an existing status report, a report on the recommendations on development and capacity projects, guidelines on production of the recommendation report, a prioritized list of development and capacity projects, guidelines and methodologies on the production of a prioritized list, and recommendations on capacity-building in planning and developing the plan (PMDFC, 2005).

Tendering, contracting, operation, and maintenance

PMDFC anticipated that improving the existing system through an upgrade of the infrastructure; training existing staff and bringing in technically competent staff; and providing grant funding could be completed in three years. The water scheme was started in Bhalwal in 2006 by the PMDFC under the Punjab Municipal Services Improvement Project. A contractor was hired through well-designed procurement rules to carry out the construction work. The contractor hired for the job happened to be the brother of a member of the national assembly of the ruling party from Lahore. It was ironic that on the one hand the CM was using everything in his power to provide water to the citizens in his province, and on the other hand, the influence of a member of his own party was used to defeat the CM's objectives. It seemed to be a no-win situation. The intent of good governance was cancelled out by the unscrupulous interests within the system. This is what Akhter Hameed Khan had likened to "anarchy." It meant that there was a state of inertia and the field was open for anyone to make the difference. The situation at hand required an intervention big enough to tilt the balance and small enough so that it could be carried out by community workers and experts. These community professionals, known as barefoot doctors in China, local resource persons in India, and paraprofessionals in conventional jargon, can tilt the balance. ASB's strategy was built on mobilizing such community resource persons to transform the failed governance system.

The failure of the governance system was reflected by the fact that the water supply scheme was dysfunctional even after the expenditure of enormous resources. The TMA started the scheme with a budget of 6.5 crore (65 million) rupees in 2006. The contractor abandoned the work after completing 45% of the job (Sahi, 2015). He got his bills cleared and did not bother to return. A new contractor was hired and completed the scheme with an enhanced budget of 13.5 crore rupees. Even the completed scheme was executed poorly due to staff corruption, in collusion with the contractor. There was no power connection in place to pump water to the users. ASB now took over the task for restoring this dysfunctional scheme.

Restoration of broken links as a business model for accessing water

According to the rules, a request for initiating a water supply scheme is submitted to the provincial Public Health Engineering Department (PHED) by local TMA or a member of the provincial or national assembly. PHED prepares cost estimates for the scheme and submits the feasibility assessment to the provincial Planning and Development Department (P&DD) as a demand for inclusion in the Annual Development Plan (ADP). After approval of the scheme PHED outsources the work to a private contractor through the tendering process and on completion hands over the scheme to TMA. TMA operates and maintains the schemes and recovers water charges from the consumers. However, most of the schemes fail soon after completion.

ASB was of the view that the failure of water services was caused by misman-agement of water demand. The first flaw in effective demand management is that TMAs charge water usage on flat rates instead of billing for the amount of water consumed by recipients. This practice is justified due to the assumption that it helps the poor. In reality, the poor pay for more than they receive and the rich con-sume more than they pay for under the fixed monthly rate. Water metering, which is perceived to be "privatization" of service delivery and make access to water out of reach of the poor, on the contrary, helps the poor. It calls for thrifty use of water and makes access to water affordable for the poor while generating enough resources for the TMA to effectively operate and maintain the service delivery. Access to water for all can be achieved by water metering and recovery of revenue based on a transparent and rational billing system. The experiment by Anjuman Samaji Behbood (ASB) – a community-based organization – in Bhalwal town is an excellent example of this approach, and it has demolished many myths about water shortages. In a short span of time it has shown impressive results.

ASB was invited by the chief minister (CM) Punjab to help TMA Bhalwal restore its dysfunctional water supply scheme and run it on sustainable basis. The pro-gramme started in December 2012. It is important to remember here that ASB did not receive any grant funding for this project from any source. They only restored the missing links throughout the governance system. The first missing link was a lack of understanding of the critical importance of interdependence between the community and government in the successful operation and maintenance of a water supply scheme. Both the government and community brought in different types of knowledge, resources, and competencies, and their mutual mistrust resulted in management failure. ASB's response was to define their respective roles and build a partnership for better service delivery. This partnership-based model was known as a component sharing model in Pakistan and was introduced by OPP to provide sanitation services to the residents of the largest informal settlement, Orangi in Karachi. ASB introduced the component sharing model for restoring the dysfunc-tional water supply scheme in Bhalwal. Under this model, the government finances the infrastructure (tube wells, overhead tanks, the disposal system, the treatment plant, and the main pipes), known as the external component; and the community

Figure 6.1 Nazir Ahmad Wattoo (second from left) and Dr. Shujaat Ali, provincial
secretary (third from left) launching the Bhalwal programme

bears the cost of purchasing lane-level water pipes, and water meters, known as
the internal component. The community pays fees based on the quantity of water
consumed (Iftikhar et al., 2018).

The second missing link was imprudent financial management and poor technical oversight of the construction work. During its survey, ASB found out that water
lines laid by the second contractor at double the cost of original project leaked at
numerous points. The quality of the pipe was also defective. The pipe manufacturing
company in collusion with the contractor and consultants had provided substandard
pipes.[3] The contractor also built the water supply infrastructure without any map or
layout of the pipelines. Government officials absolved themselves of any responsibility by stating that under procurement rules they were not allowed to negotiate
with the pipe manufacturer. The community became aware of the problems and
ASB mobilized the public, and the pipe manufacturer agreed to replace 975 metres
of pipes made of substandard material.[4] This helped get rid of the leakage problem.

The third missing link was the lack of a support organization to guide the community to play its role in internal development and keeping checks and balances
on the government's work. To create this link ASB signed a Memorandum of
Understanding (MOU) with the administrator and chief officer of TMA for the
restoration and efficient operation of the water supply scheme on December 31,

2012. ASB was given office space on the TMA premises to establish and run a Water and Sanitation Committee (WASCO). ASB planned to provide water 24/7 to 4,000 households (Sahi, 2015). ASB created a two- to three-person team for diagnostic work. The staff was hired by the TMA as daily wagers under its contingency budget. This significantly reduced TMA's budget for diagnostic work. These workers received Rs. 10,000, compared to Rs. 35,000 paid to lowest level of government employees. Diagnostic work consisted of mapping and digitization of the physical infrastructure and holding dialogues with the community members for water metering. In addition, a survey was carried out for house numbering and collecting demographic data, and a pedestrian survey was conducted to identify leaks in the pipes and joints. ASB also arranged meetings between community and government officials at various levels to clearly identify the complaints, the required measures to address critical issues, and the roles to be played by different partners. The meetings also functioned to build trust between the community and the government.

In 2013 ASB got a power connection and started pumping and distributing water. During the course of its work ASB identified 1,000 cases of water leakages showing poor quality of workmanship. At the same time a Water and Sanitation Committee of community notables was formed to help ASB in its diagnostic work and community dialogue. The committee showed serious limitations. It was gradually phased out and a new committee consisting of local activists was formed to take over the assigned tasks. When ASB took over the responsibility for collection of charges for operation and maintenance of the water supply scheme, the recovery rate was 50%. Due to water metering, water consumption decreased by more than 50% and water charges recovery increased to 90%. ASB introduced a tariff of Rs. 50 per 1,000 gallons, Rs. 70 for an additional 1,000 gallons, and increased it to Rs. 100 for the next 1,000 gallons. As a connection holder, each household had to pay a fixed charge of Rs. 50 per month (Iftikhar et al., 2018). This resulted in recovery rates of more than 90% compared with traditional models having less than 50% (Iftikhar, 2016)

At present 2,000 out of 2,500 households in the project area have received metered connections and ASB recovers 95% of water charges amounting to Rs. 700,000. This is in contrast to the previous practice of levying Rs. 30 per household in water charges and experiencing a high default rate. ASB has succeeded in reaching a high level of recovery due to effective and efficient service delivery; this has enabled ASB to pay the salaries of 16 employees hired by the project and has made the scheme fully functional.

Conclusion

The answer to the dysfunctional services comes down to making a choice between more spending and better spending. ASB's business model provides a brilliant example of better spending, and it depends on the leadership role played by a support organization that builds trust between the community and the local government by restoring the broken links through technical and social guidance. One can say that ASB's success very much depended on the special circumstances, the personal commitment of the chief minister, and the presence of a key ally in the civil service. This special situation is not in violation of the general principle that you build linkages though finding champions in every constituency. The process of transformation does not take place by overlaying a new governance system over an old one but by triggering the process of change by identifying and engaging the catalysts in the community and the government.

Notes

1 DCO functioned as the head of the district administration and replaced the former deputy commissioner during the devolution of power to the local government programme introduced during the tenure of General Pervez Musharraf (1999–2012).
2 FB study conducted for WSP of the World Bank to draw lessons from the successes and failures of WUC in Punjab.
3 Based on a telephone interview with Mr. Nazir Ahmad Wattoo, CEO of Anjuman Samaji Behbood, on November 26 and 27, 2018.
4 Ibid.

References

Iftikhar, M. N. (2016) Community-driven: Bhalwal contributes viable business model of drinking water, *Express Tribune*, 30 May, Business Section.
Iftikhar, M. N., Ali, S., & Sarzynski, A. (2018) Community – government partnership for metered clean drinking water: A case study of Bhalwal, Pakistan, in Hughes, S., Chu, E., & Mason, S. (eds.), *Climate Change in Cities*. Cham, Springer, pp. 163–197.
PMDFC (Punjab Municipal Development Fund Company). (2005) *Field Appraisal Report TMA Bhalwal*. Lahore, PMDFC.
Sahi, A. (2015) *Water in the* pipeline, in Action Aid (ed.), *Water, Sanitation and Hygiene in the Media: A Collection of Stories by Journalists of Media Fellowship Programme.* Islamabad, Action Aid.
Tahir, M. A., Akram, C. M., ul Hasan, E. F., & Farooque, E. M. (2011) *Report on Technical Assessment of Water Supply Schemes Northern and Central Punjab*. Islamabad, Pakistan Council for Research on Water Resources (PCRWR).

7 Redefining and localizing development in Pakistan

Furqan Asif

Pakistan has received billions of dollars in international aid and loans to support development in the country. Much of these funds have not achieved the stated social and economic goals set out by the projects, a conclusion reached by independent evaluation units within the very international financial institutions (IFIs) giving the funding. The reason for this is largely because projects fail to consider local realities and contexts and lack input from local communities. Amidst the squandering of millions of dollars and ineffectual urban development, local organizations have emerged as a strong voice for pro-poor development. This chapter highlights two such organizations in Karachi, the Urban Resource Centre (URC) and the Orangi Pilot Project (OPP), to showcase how local communities can use their own resources combined with technical assistance and lobbying/advocacy via the media to question IFI-backed government projects while proposing alternatives. Through a case study involving the Karachi Wastewater Management Project, I show how the URC and the OPP worked together by collecting documentation, mapping out existing local development, commissioning expert analysis, and mobilizing advocacy efforts targeting all levels of government to support an alternative plan, which ultimately led to the cancellation of a US$70 million loan from the Asian Development Bank. Examples like this empower local communities, paving the way for redefining development that is "by the people and for the people."

Keywords

Participatory development; Karachi; Pakistan; aid effectiveness; urban planning; Urban Resource Centre

Introduction

Pakistan has received approximately US$170 billion dollars in international funding/aid from 1952–2013, including US$18.4 billion from 2010–2013 (AidData, 2019). How has this impacted development? While there is an ostensible connection between aid and improvements in development outcomes (e.g., poverty reduction) (Newman, 2013), there are two points that complicate this picture. One is that the primary driver of recent poverty reduction has not been development aid per say, but rather, the growth in income – specifically led by movements of the labour force out of agriculture and into industry and services (Inchauste & Winkler, 2012). This shift has been supported by policies and investments that boosted productivity in the non-agriculture sector, leading to modest pro-poor growth favouring the poorest 40% of the population (Lopez-Calix et al., 2014). Non-labour transfers (e.g., international remittances, Zakat/Usher) and social protection initiatives have also played a critical role in broader poverty reduction interventions, while empowering women (e.g., via transfers money to the female household head) (Lopez-Calix et al., 2014; World Bank, 2013). Altogether, the outcome has been that the share of people living below the poverty line has been cut in half from around 50% in 2005 to 24% in 2016 (Ministry of Finance, 2017, 245).[1]

The second point is while foreign aid has been substantial, it has not been as effective and/or well spent, largely due to poor planning, misplaced incentive structures, and a failure to understand the way social and institutional norms prevent progress from being made (Fried & Lawrence, 2003; Hasan, 2010). Somewhat ironically, these conclusions have been supported by the independent evaluation units of two major donors to Pakistan, the World Bank (WB) and the Asian Development Bank (ADB). The Independent Evaluation Group (IEG) rated the outcome of the assistance provided by the World Bank Group (WBG) to Pakistan from 2010–2014 as "moderately unsatisfactory," but with some areas such as "improving human development and social protection" as "moderately satisfactory" (IEG, 2014, 13). Meanwhile, the Independent Evaluation Department of ADB concluded that for the 2002–2012 period, "the performance of ADB-supported operations has been poor" partly attributed to changes in the modality and sector focus of ADB's portfolio after 2001 (ADB, 2013, v).

At the same time, talk of aid effectiveness and poverty reduction is largely dominated by a focus on bilateral aid agencies, development banks, and international NGOs (no doubt given their deep pockets and the large sums involved). Amidst this backdrop, it can be easy to overlook the fact that poverty reduction happens at the local level and that what *actually* happens "on the ground," versus what is written in a terms of reference (ToR) or project proposal, is what ultimately makes the difference. By extension, often the challenges to poverty reduction are local – that is, local power structures, land owning/titling systems, anti-poor politicians, bureaucracies, and regulations (Hasan, 2007). Further compounding this, many of the resources and much of the autonomy that are needed at the local government level are often not transferred and remain at higher (provincial) levels (MHHDC, 2014, 118). For example, provincial legislation in Punjab (Pakistan's most populous province) dictates that local governments are responsible for a bevy of areas: link roads, intra-urban roads, street lighting, solid waste management, fire response, parks, playgrounds, curative health, primary education, water supply, drainage, and sewage. Yet, the allocations of resources/funds are assigned to the province (Shaheen, 2018). Such a disconnect between roles and responsibilities; uncoordinated arrangements between local, provincial/state, and national government; and the lack of resource decentralization for large municipalities in Pakistan has greatly hindered public service delivery (MHHDC, 2014).[2]

In such a chaotic institutional context, many needs of the poor (e.g., schools, healthcare, water and sanitation, land, social safety nets) are being fulfilled by local community-based organizations (CBOs) (e.g., local NGOs, grass-roots organizations formed by the poor), often with the support of local governments (Hasan, 2010). Despite operating on relatively small budgets and outside the major funding flows and official development assistance, such organizations have a unique understanding and knowledge of the local context and are accountable to the local population. The impetus to organize locally and act has also been a direct response to seeing the inefficiency of donor-led development efforts, which are often controlled by a myriad of actors: uninformed and rent-seeking politicians, private interests, loan-pushing international finance institutions (IFIs), opportunistic national and international consultants, and profit-seeking local and foreign contractors and companies (Hasan, 2010).

One such organization is the Urban Resources Centre (URC) in Karachi. In the 1980s, the ineffectiveness and failure of large-scale (in scale and funds allocated) projects was noticed by a group of professors at the Department of Architecture and Planning in Dawood College, along with citizen groups. The professors used their position as educators and put forward a Comprehensive Environmental Design Project with the aim of having students understand the broader social, administrative, and economic issues in the urban landscape of Karachi. The project became a permanent fixture within the department and each year a different area of the city that faced social/environmental challenges was chosen. Students were divided into four groups (physical, administrative, social, and economic) and each researched and observed the problems, determined local government responsibilities, identified interest groups and local organizations, and homed in on the root of the issues faced by the community. Each student had to suggest a specific architectural intervention for improving the area with the whole group synthesizing their findings and arriving at solutions for the various areas. A consequence of the annual project was that, over time, it created strong links between students/teachers and inner-city interest groups, the informal settlements (locally known as *katchi abadis*), informal sector operators, NGOs, and government departments (Hasan, 2007).

Among the links, some were institutionalized, such as with the Orangi Pilot Project (OPP) institutions[3] that were working in the katchi abadis. Almost in parallel to the beginnings of the URC, the OPP was established in 1980 through a Pakistani charity (the Bank of Commerce and Credit International [BCCI] Foundation), led by its chairman, Agha Hasan Abedi, and Dr. Akhtar Hameed Khan, a nationally renowned development practitioner. The organization's purpose was to develop community participation and local resource mobilization in response to problematic, ineffective government programmes set up to upgrade poor informal settlements and address poverty alleviation, with a focus on the Orangi Township (Hasan, 2010).

In 1989, the Urban Studies Forum was established within Dawood College. The forum collected information and carried out investigations into the key urban challenges in Karachi and disseminated the results. At the same time, forum members noticed how Karachi's official development plans often ignored or overlooked the broader socio-economic and environmental landscape of the city, thereby making the plans untenable. Thus, it became clear that in addition to their existing activities, they also needed to provide a counter response and challenge to the ineffective and failed donor-funded projects. As such, the Urban Resources Centre was established as a research-focused documentation and advocacy organization focusing on urban challenges in Karachi (Hasan, 2007). The organization consisted largely of professionals in Karachi: teachers and young graduates at Dawood College; activists from the Orangi Pilot Project (OPP); urban planning professionals; and members of community organizations from low-income settlements. The URC wanted to improve the urban planning process for Karachi through involvement of local, informed communities and interest groups, and in doing so, ensure the planning processes would serve the interests of the low- and lower-middle-income groups and empower the city's citizens to take ownership of development (Hasan, 2010). In practice, the organization acts as a hub of a network of NGOs in Karachi and engages in public advocacy work (Sansom, 2011).

The organization has sought to achieve its goal through three areas of activity. The first is regularly monitoring and reviewing news media to identify new projects under government consideration (including tracking budgetary process and financial allocations for projects under review). Once identified, the URC conducts its own technical and financial review of the government's project document (known as PC-1) using its membership base of urban planning professionals and activists. After the analysis, it arranges a public hearing of the reviewed proposal with various interest groups (e.g., informal service providers, government agencies, political parties). In cases where there is an inappropriate allocation of resources, the organization organizes petitioning activities to demand budgetary review of the project. The second is through mapping exercises by the local community and involving community-based organizations (CBOs), NGOs, interest groups, charities, and local action groups. This involved mapping government departments and collecting their contact information and standard operating procedures for participation in the technical review of the proposed projects. The third activity area involves knowledge-gathering/generation and advocacy through collecting documentation (e.g., tenders, ToRs); commissioning expert analysis; providing participatory alternatives to conventional development plans (e.g., holding meetings and bringing key government officials into discussions); building community awareness; and mobilizing pressure through media. Altogether, these activities have created a network of professionals, activists from civil society, local groups, and government agencies who have a localized understanding of planning issues/challenges, which has actively challenged several government plans that are inefficient/ineffective, exorbitantly expensive, and anti-poor, and promoted alternatives (Hasan, 2007).

Through a case involving a donor-backed sewage disposal plan for Karachi, this chapter shows how the URC has helped to redefine and take ownership of development by supporting local CBOs, while challenging the dominance and "expertise" of major donors. It underscores the importance of better spending rather than more spending and how local organizations and actors play a critical role in enhancing human development through the use of local expertise/actors and knowledge-gathering/creation to lobby for alternative, just, and sustainable development planning/programming. In doing so, the hope is that this story offers a model and inspiration for other countries and contexts on how to contribute to Sustainable Development Goals through localizing development.

The urban context and the katchi abadis

Development within Pakistan is often synonymous with "the urban," and it is easy to see why: 37% of the country's population of 207 million live in urban areas and the country is urbanizing at an average of 2.91% (1990–2015) (Rana, 2017; UN DESA, 2018). Approximately 20% of this portion (~15 million) live in the country's most populous (and only) port city, Karachi, located in Sindh province.[4] The city's population has increased 115% since the previous census in 1998, and it is the third most dense city in South Asia (Ferrer, 2015). Such rapid urbanization has brought with it several challenges related to land use and transportation planning, infrastructure (sewage and wastewater), and the proliferation of high-density informal settlements in the peri-urban area, locally known as katchi abadis. From an urban planning and development point of view, the katchi abadis are of particular importance because they are where over half of the city's population live, and the overwhelming majority of those in the katchi abadis live below the poverty line (Hasan, 2007). Given their informal nature, the katchi abadis have traditionally been plagued with inadequate provisioning of basic services such as housing, water supply, and sewerage systems.

While there are underground sewage systems, they have not kept pace with urban expansion and have been inadequately maintained. As such, much of this infrastructure has collapsed in certain areas, with ad hoc rehabilitation measures taken to connect them to natural drains or water bodies (Pervaiz et al., 2008). In the absence of a citywide disposal infrastructure, much of the waste generated in the informal settlements has been collected via bucket latrines or soak pits and disposed of via open drains that run into the natural drainage system, water bodies, or natural depressions (Ferrer, 2015). Overall, the sewage system only services 40% of Karachi's population; however, most of the sewage and wastewater is released into the sea untreated (Pervaiz et al., 2008). As a result, disposal of untreated sewage, particularly industrial waste, into the sea has led to significant pollution of Karachi's shoreline, thereby having a detrimental impact on coastal and marine life, which has consequently aggravated the lives and livelihoods of fisherfolk (Yousafzai, 2019).

The Orangi Pilot Project and its low-cost sanitation programme[5]

The first step in improving the sanitation situation involved Dr. Khan visiting Orangi Township and talking to its residents – shopkeepers, tea house patrons, people working in the informal sector, and community interest groups – to better understand the community. Through these conversations, several community leaders/activists were identified and subsequently, space was rented in a modest commercial building to establish the OPP office (Hasan, 2008a). Through dialogues with the Orangi community, it became clear to Dr. Khan that both residents and interest groups were not interested in locally led development. Instead, many community organizations vied for patronage from political parties and the local bureaucracy and lobbied for micro-level improvements in their neighbourhood. In turn, the organizations offered votes and support to these political parties. While people did receive some benefits, they also ended up becoming dependent on them, thereby only reinforcing and entrenching this patronage system (Hasan, 2010).

Notwithstanding such a system where only a few benefitted from such a neo-patrimonial relationship, people in Orangi were very keen on improving their lives by building schools, addressing inadequate water and sanitation, and improving their health and environment. However, Dr. Khan observed that the lack of success was not due to community will; instead it was due to not having access to technical assistance, managerial guidance, credit, and information on government support programmes, which in turn, resulted in financial resources being used ineffectually (Hasan, 2010). Over a six-month period of dialogue and investigation in the Orangi community, Dr. Khan identified four major areas people faced challenges in that, if addressed, would improve their life conditions (in order of their priority): sanitation, health, education, and employment (improving income).[6]

Sanitation was identified as the first area to be addressed, given the detrimental impacts a lack of sanitation was having on the health and well-being of Orangi residents. When the OPP office was established in 1980, the lanes (alleyways) of Orangi were flowing with wastewater and sewage from the houses. The residents would use bucket latrines, which were serviced by someone who would empty them out every fourth or fifth day at a cost of US$0.25 per month. The more affluent households had built soak pits, but these were plagued with problems (most had filled up and were not functioning properly because of shoddy construction), while others put in sewage lines from their house to the nearest natural drain (but these were often defective). It became apparent that an underground sewage system needed to be constructed. Yet, the residents of Orangi expected that such a system was to be provided for free by the local government or Karachi Development Authority (KDA), and this mentality was reinforced by the local leadership, who supported this mindset. When contacted by Dr. Khan, the KDA, in turn, noted that they did not offer anyone a free sanitation system and the communities that did have it had paid for it based on a user fee, an amount that was far out of reach of Orangi residents. They suggested that the need for sanitation could be provided through the Katchi Abadi Improvement Programme (KIP), which started in 1978 and was financed through international loans by the World Bank and ADB. The programme gave a 99-year lease to the residents of katchi abadis through payment of a development charge. However, the initiative was beset with problems and had not been successful because not only did it have complicated procedures and lacked transparency and accountability, but it also suffered from a lack of community involvement/participation and exacerbated the mistrust between communities and local government organizations. On top of this, the regularization process (i.e., formalizing the informal settlements) was painfully slow: by the 1990s, it was at a pace of only 1% of katchi abadis per year, which meant that it would take 100 years for the process be completed (Hasan, 2010).

During subsequent discussions with the KDA and Karachi Metropolitan Corporation (KMC) and probing as to why the cost of the sanitation system was so cost prohibitive to the katchi abadi residents, Dr. Khan discovered that the billed cost by the KDA and KMC was *seven times* the true cost when considering the labour and materials involved in constructing an underground sewage system. The cost inflation was especially pronounced when foreign financial assistance was involved and, in such cases, the base costs increased by 30 to 50%, largely due to fees of foreign consultants, and in cases where international tenders (or bids) were involved, costs increased 250%. Thus, it was established that the actual cost of an underground sewage system was likely not "expensive," especially if it involved the local community (Hasan, 2010).

To confirm this, the first step involved setting up a technical unit within the OPP, which included an Orangi-based engineer alongside a local plumber, a draftsman, and a technician. They were provided with survey and levelling instruments to support the mapping out of design of the underground sewage system. The second step focused on establishing community organizations which consisted of people from a lane (i.e., a row of 20 to 30 houses).[7] In order to have community-wide buy-in, a campaign of awareness-raising by Orangi residents and activists, supported by the OPP, was conducted. The campaign consisted of holding meetings with lane residents to explain to them (with the aid of slides, models, and pamphlets) the connection between their children's illness and the poor sanitation environment, along with the economic costs incurred as a result (i.e., lost wages for time off and cost of medicine). They also countered misinformation by telling them that neither the KDA or KMC installed sewage lines for free and that the charge was cost prohibitive for Orangi residents. On top of this, they informed residents that it would likely be years before a sanitation system would come to Orangi through government programmes (Hasan, 2010).

The Orangi social organizers informed the people that if they formed an organization with residents from their lanes where they participated, elected, or nominated lane managers, the OPP would assist them. Two managers would be selected, one to manage the purchase of materials and organize the work, and the other to manage money and accounts. This consisted of the OPP technical staff surveying the lane, establishing benchmarks, and preparing plans and estimates (labour and materials), with this data being given to the lane managers. Then, the lane managers would collect the money required from the residents and arrange meetings to resolve any technical or social challenges (OPP staff supervised the process but at no point did they handle the money). In this way, leadership and community formed at the lane level. While the initial phase was not without its challenges (e.g., substandard work; clogged drains; concrete weathering rapidly), after an evaluation was done and issues identified along with solutions provided, the programme was improved. The programme grew, supported by training programmes (e.g., for local masons) and many lane organizations came together, collecting money and obtaining technical input from the OPP – creating a confederation of lanes. In 1990, the average cost of a sanitary latrine for a house (primary and secondary drain) was about US$16.66 per household, a figure affordable by most Orangi residents (Hasan, 2007).

The low-cost sanitation programme enabled low-income households to construct and maintain an underground sewage system using their own funds, with the system staying under their management. The OPP-Research and Training Institute (OPP-RTI) provided social/technical guidance (informed by action research), tools, and supervision of the implementation. Part of this support also included developing new construction standards and techniques that were affordable to the local community and in line with their vision of their involvement in the construction. Most importantly, the people in the community collect, use, and manage funds themselves, with the institution providing support as required or requested. Such a model showed that local people have the ability and means to finance and build underground sanitation systems for their homes, lanes (alleys), and neighbourhood – this was subsequently termed "internal" development by OPP-RTI (Hasan, 2010). At the same time, it was unreasonable to expect people to build trunk sewers (i.e., the "main"' line connecting to the secondary lines), treatment plants, and larger secondary sewers. Such large-scale "external" development is something only the state can provide. Such a state agency-community partnership came to be known as the *collective sharing model*, with internal development constituting 70% of the total cost of the sewage system and external development 30%. In Orangi, over 95,000 households have built their neighbourhood sanitation systems by investing US$1.4 million. By contrast, for the same work, the local government would have had to invest US$10.5 million (OPP-RTI, 2004 as cited in Hasan, 2006).

Role of the URC and redefining development

Throughout this period as the OPP-RTI was supporting a locally led sanitation model in Organi, the URC was being informed of its success from the network of CBOs and local community organizations via a forum it had established that brought them together to discuss problems, learn from each other, and take action to collectively protect the interests of the community. Seeing the power and efficiency of true participatory local development, the URC started to question several government urban development projects and began proposing pro-environment, pro-poor alternatives. To do this, it launched an aggressive advocacy campaign by liaising with media along with establishing links with the Asian Coalition for Housing Rights and UN agencies (Hasan, 2007). While this created some conflict of opinions between URC and local government and its planners, by the late 1990s and early 2000s, relations improved, no doubt after local development projects such as the low-cost sanitation programme supported by OPP-RTI were bearing the fruits of their success. The URC presented findings from programmes on solid waste, transit, and inner-city issues conducted by CBOs to provincial governors, ministers, and local government officials including Karachi's mayor. At the same time, the URC played a crucial role in bringing together and supporting civil society mobilization against international loans and mega projects (Hasan, 2007).

One of the earliest successes where URC's role as an advocacy and knowledge-gathering/dissemination organization came to the fore was in the late 1990s when it began supporting OPP-RTI's questioning of an ADB-funded Greater Karachi Sewage Plan (GKSP). The Karachi Water and Sewage Board's (KWSB) GKSP was engineered to provide both "internal" and "external" development; however, the project had been unsuccessful. The lack of success was determined by OPP-RTI through its mapping activities, which showed that the GSKP ignored the existing sewage systems that were already in place and discharging into the natural drainage areas of the city. Partly, the project overlooked the pre-existing system because the work done was considered "substandard" by consultants hired by ADB. As such, by only building trunks along the main road and not incorporating the existing sewage lines, the treatments plants were only functioning at 25% of their capacity (Hasan, 2010). OPP-RTI prepared and published a comprehensive report ("Proposal for a Sewage Disposal System for Karachi"), but KWSB planners and engineers objected the proposed recommendations citing that they were against good engineering practices (e.g., sewage and rainwater flowing together – something that OPP-RTI showed was a practice followed in countries like Japan). To make matters worse, the weight of the US$700 million ADB loan was extremely burdensome for the KWSB, putting a strain on the economy of the city and province (Hasan, 2010).

In this context, the URC, with the endorsement of OPP-RTI, helped to promote and share the alternative local low-cost sanitation model championed by OPP-RTI among NGOs, CBOs, academia, and government agencies by arranging meetings, forums, and lectures. Meanwhile, the OPP-RTI was also proposing alternatives for another large-scale sanitation project, the Korangi[8] Wastewater Management Project (KWWMP), which was being financed by a US$70 million loan from the ADB with US$30 million being provided by the provincial government. Engaging with community activists in Korangi, OPP-RTI determined that most lanes in the neighbourhood had been built by the community, but this development was being ignored by the project, and instead of being integrated, would be redone/replaced. In 1998, the OPP-RTI gave a series of presentations of its alternative plan to the KWSB, the provincial government departments, the Planning Commission in Islamabad (the country's capital), the president of Pakistan, the provincial governor, and the ADB. After a series of discussions and debates, in April 1999, the provincial governor decided to cancel the US$70 million ADB loan, marking the first time in the history of Pakistan that an IFI loan had been cancelled (Hasan, 2010; Sansom, 2011). A committee was formed at the request of the governor to develop a plan; it was drafted by OPP-RTI in March 2000. By incorporating the knowledge of the local context and conditions – for example, complimenting the existing system and using local resources and technology/knowhow – it showed that the cost of the project would be US$15.18 million (Hasan, 2010). Ultimately, the plan was not implemented, but the case of KWWMP and subsequent cancellation of the ADB loan generated considerable discussion and debate in the press and among the CBOs/NGOs. In turn, this led to the formation of a coalition of 59 NGOs and CBOs (including OPP-RTI) that advocated (via a position paper published widely) for a community-based agenda on Karachi's water and sanitation. Last, and perhaps most important, this series of events catalyzed an anti-IFI movement not only in Karachi but also in the country overall (Hasan, 2010). These days, sanitation is chiefly funded via grants from the provincial and federal government (Sansom, 2011) – a consequence no doubt connected to past experiences and lessons learned.

Discussion

While the federal, provincial, and city governments have invested significantly in Karachi's development efforts, this has often been supported via loans from IFIs. However, many projects either fail to achieve the stated project objectives or have been shown to be unsustainable. This is the conclusion from a report that analyzed data from ADB's own audit documents for Pakistan: "Perhaps as many as 70 percent of ADB projects in Pakistan are unlikely to produce lasting economic of social benefits" (Fried & Lawrence, 2003, 2). Since 1960, Pakistan has received over US$12.6 billion in loans from the ADB, making it the second largest borrower (after Indonesia) in 2002 (Fried & Lawrence, 2003). Notably, the reasons identified include design flaws and lack of attention through project preparation, which included a lack of a benefit and monitoring system and baseline data needed to measure successes, and a consistent lack of consultation with beneficiaries and local groups, as well as and no meaningful level of community participation. Moreover, once the project is completed, there is a debt burden, exacerbated by the high interest rates, that the government must service. When you consider, for example, that the government collects only 9% of its GDP in taxation revenue[9] (Baqir, 2019) and the country ranks 150 out of 189 on the Human Development Index (HDI) (UNDP, 2018), it is easy to see the challenge of paying off such loans while endogenously financing overall development within the country. However, despite such a precedent of inefficiency and failure, the government has continued to negotiate new loans for large-scale projects (Hasan, 2010). At the same time, there is no denying the need for investments and projects to address water supply, sewage, effluent treatment plants, and so on in Karachi and other parts of Pakistan. What is contested is the model by which such areas are addressed, with many professionals and academics, and a number of NGOs (including URC) feeling that Pakistanis do not need loans or foreign experts to execute these projects nor are all projects necessarily appropriate or needed (Hasan, 2007).

These issues are set within the broader challenges that stymie development, one of the most glaring being the fact that Pakistan lacks a comprehensive urban policy framework (MHHDC, 2014). While the Planning Commission did set up a task force on urban development in 2010, which laid out solutions to address the negative consequences of urbanization, these solutions have not been systematically implemented. Karachi has been of central focus because uncontrolled urban sprawl and haphazard development has led to uneven infrastructure provisioning and a polluted urban environment, combined with a paucity of local engagement. Even though the city has developed five master plans since 1923, none have been effectively implemented (MHHDC, 2014).

Notwithstanding these wider, systemic challenges, local CBOs have played a critical role in both initiating and advocating for development on the terms of the members within local communities and in their interest – exemplified by the KWWMP story. If anything, the case of localizing development in Pakistan shows the success that can be achieved if local people realize the power of organizing and accessing technical assistance through pioneering organizations such as the URC, despite the broader challenges. Moreover, there is a need to create an information base that counters the prevailing narrative and assumptions, which can be done through the development of a repository/library of news clippings, publications, and key documents (e.g., IFI proposals, local government reports) as showcased by efforts of URC. Using that information and disseminating it through the media (e.g., letter writing campaigns) leads to wider coverage, more awareness, and participation by citizens, which in turn leads to issues landing on the radar of political parties (who curry favour from their community), academics, community activists, and public sector organizations.

Another dimension for establishing strong local organizations and networks that support pro-poor development is the building of trust. Part of the initial success of OPP can be attributed to the social foundations and trust that was established at the outset. The founder of OPP had respect within the corridors of the government bureaucracy because of his prior tenure as a member of the civil service, which in turn allowed him/OPP to obtain easy access to government officials. While this strategic use of former connections certainly helped at the outset, the relationship with the government was developed gradually over several years with ongoing efforts by management within OPP to foster the relationship. The institutionalization of this culture of placing an importance on relationship building was an essential factor in getting government officials to sit at the table and come to the meetings that OPP organized to discuss its projects and solicit input from the community. Government engagement was also catalyzed, with support of efforts by URC, after OPP was able to demonstrate the success of the low-cost sanitation programme and showcase its role in building links within the Orangi community (Pervaiz et al., 2008; Sansom, 2011).

Conclusion

For Pakistan, as is the case with many other developing countries, the vision of urban planning and development by the government invariably follows the "Western" model – not because it is the best or ideal, but because of the hegemonic influence of IFIs alongside political and economic opportunism, from both international and domestic actors. None of these dynamics and players operate at the local level, the scale at which development occurs. The consequence of this can be seen across the urban landscape of countries like Pakistan: high-rise apartments and condominium complexes instead of upgraded informal settlements;

large shopping malls instead of traditional local markets; massive expressway projects instead of traffic management and planning; gentrification and pushing poverty to the urban periphery instead of developing affordable housing (Hasan, 2007). Such development tactics serve neither the interests of the poor nor the environment – indeed, they ultimately do not serve the interests of most of the population (save perhaps the elite/affluent class).

However, if people are brought together and form local organizations and gain a better understanding of the challenges and solutions, they can challenge the dominant urban planning models and projects (Hasan, 2008b). Questioning and critiquing government initiatives that are not pro-poor or those that do not consider the reality of local communities in an informed and strategic way that is backed by a coalition of actors (CBOs, NGOs, academics) can force the government to listen and, eventually, change its initial plans/projects/investments. Such victories accrue over time and validate/empower local communities, paving the way for redefining development that is "by the people and for the people."

Notes

1   Getting an accurate indication of poverty indices and reduction is murky given that Pakistan has its own national poverty line, which differs from conventional poverty metrics insofar as it is adjusted for inflation using the Consumer Price Index (CPI); however, the government does not collect CPI for rural areas, which means there is no singular poverty line for both urban and rural areas.
2   Yet another challenge is that, at the national level, Pakistan's urban centres receive insufficient funding and policy attention because the national government prioritizes other areas such as debt repayment and military spending (MHHDC, 2014).
3   The OPP is the designation of three Pakistani NGOs (OPP-Research and Training Institute; OPP-Orangi Charitable Trust; and OPP-Karachi Health and Social Development Association) that formed and emerged from the original organization, which conducted innovative community-based social programs in the katchi abadis of Orangi Town in Karachi.
4   There has been ongoing controversy surrounding the accuracy of the population numbers of Karachi (DunyaNews, 2017), with the provincial assembly refusing to accept the 2017 census figures for the province, citing irregularities such as the fact that millions of people were listed under multiple addresses and counted in their home province despite living in Sindh (Ghori, 2017). According to the National Database & Registration Authority (NADRA), which was not consulted in the census, the population of Karachi is 20.15 million. On April 25, 2018, Mustafa Kamal, chairman of the Pak Sarzameen Party and former mayor of Karachi, petitioned the Supreme Court and requested that a third-party audit of the national census be conducted (Arshad, 2018).
5   Much of this section is the result of a condensed synthesis, adapted from Hasan (2010, Chapter 6), who outlines the beginnings and evolution of the low-cost sanitation programme of the OPP.
6   These four main issues were addressed through its five independent institutions set up in 1988: the OPP-Research and Training Institute (RTI) focusing on sanitation, housing, education, training, research, documentation, and advocacy; the Orangi Charitable Trust (OCT), which operates a microcredit programme; the Karachi Health and Social Development Association (KHASDA), focusing on the operation of a health programme; the Rural Development Foundation, which conducts agriculture-related research and extension (i.e. application of research to agricultural practices through farmer training/education); and the

OPP Society, which mobilizes resources from the BCCI Foundation to these organizations (Hasan, 2007).

7 There were also two other benefits with this kind of organization. One is that the unit was small and thus, easy to manage, while everyone knew each other, being neighbours, so there was no issue of mistrust. Second, the Orangi leadership did not feel threatened by the community organizations since they are small and confined to one-lane units.

8 The name of a district located in Landhi Town in the eastern part of Karachi.

9 The International Monetary Fund (IMF) notes that for a country to finance the achievement of development goals, it should collect 20% of its national income via taxes (by contrast, a developed country collects 40% of GDP in taxes) (Akitoby, 2018).

References

ADB. (2013) Country assistance program evaluation (CAPE) – Pakistan: 2002–2012. Available at: www.adb.org/sites/default/files/evaluation-document/36106/files/cape-pakistan.pdf

AidData. (2019) Aid project list: Pakistan. Available at: http://dashboard.aiddata.org/#/advanced/project-list

Akitoby, B. (2018) Improving tax collection, raising tax revenue and lessons in tax reform – IMF F&D magazine, *Finance & Development*, 55(1), pp. 19–21. Available at: www.imf.org/external/pubs/ft/fandd/2018/03/akitoby.htm

Arshad, U. (2018) Mustafa Kamal challenges census 2017 results in SC, *Daily Pakistan Global*, 25 April. Available at: https://en.dailypakistan.com.pk/pakistan/mustafa-kamal-challenges-census-2017-results-in-sc/

Baqir, F. (2019) *Poverty Alleviation and Poverty of Aid: Pakistan*. London, U.K., Routledge, Taylor & Francis Group.

DunyaNews. (2017) Initial estimates after Census 2017 put population at 21–22 crores – Pakistan – Dunya news, *Dunyanews.Tv*. Available at: http://dunyanews.tv/en/Pakistan/390117-Initial-estimates-after-Census-2017-put-population

Ferrer, M. (2015) *Karachi: City of Woes*. Available at: https://www.academia.edu/13184139/Karachi_City_of_Woes

Fried, S. G., & Lawrence, S. (2003) The Asian development bank: In its own words – an analysis of project audit reports for Indonesia, Pakistan, and Sri Lanka. Available at: https://docs.wixstatic.com/ugd/898604_678c2a067a1048efa1ac992aaf2f02ae.pdf

Ghori, H. K. (2017) Sindh Assembly refuses to accept census results, *Dawn*, 3 November. Available at: www.dawn.com/news/1367985

Hasan, A. (2006) Orangi pilot project: The expansion of work beyond Orangi and the mapping of informal settlements and infrastructure. *Environment and Urbanization*, 18(2), pp. 451–480. https://doi.org/10.1177/0956247806069626

Hasan, A. (2007) The urban resource centre, Karachi. *Environment and Urbanization*, 19(1), 275–292. https://doi.org/10.1177/0956247807076921

Hasan, A. (2008a) Financing the sanitation programme of the Orangi pilot project – research and training institute in Pakistan. *Environment and Urbanization*, 20(1), pp. 109–119. https://doi.org/10.1177/0956247808089151

Hasan, A. (2008b) *The Roles of Local Organisations in Poverty Reduction and Environmental Management*. Karachi, The Urban Resource Centre. Available at: http://urckarachi.org/downloads/URC%20Case%20Study%20by%20IIED%20(For%20Web).pdf

Hasan, A. (2010) *Participatory Development: The Story of the Orangi Pilot Project-Research and Training Institute and the Urban Resource Centre, Karachi, Pakistan*. Oxford, Oxford University Press.

IEG. (2014) *CPSCR Review: Pakistan (2010–2014)*. Washington, DC, World Bank. Available at: http://documents.worldbank.org/curated/en/964421468087559952/pdf/877150 CASCR0P0010Box385189B00OUO090.pdf

Inchauste, G., & Winkler, H. (2012) Decomposing distributional changes in Pakistan (No. 86248, pp. 1–44). *The World Bank website*. Available at: http://documents.worldbank.org/curated/en/332751468286263415/Decomposing-distributional-changes-in-Pakistan

Lopez-Calix, J., Mejia, C., Newhouse, D., & Sobrado, C. (2014) *Pakistan Poverty Trends, Scenarios and Drivers*. World Bank Policy Paper Series on Pakistan. PK 23/12. Washington, DC, World Bank. Available at: http://documents.worldbank.org/curated/en/895401468145474740/Pakistan-poverty-trends-scenarios-and-drivers

MHHDC. (2014) *Human Development in South Asia – Urbanization: Challenges and Opportunities*. Lahore, Pakistan, Mahbub ul Haq Human Development Centre.

Ministry of Finance. (2017) *Pakistan Economic Survey: 2017–18*. Islamabad, Finance Division, Government of Pakistan. Available at: http://finance.gov.pk/survey/chapters_18/Economic_Survey_2017_18.pdf

Newman, J. (2013) *Recovering Strong Positive Trends in Poverty and Opportunity*. Washington, DC, World Bank. Available at: http://documents.worldbank.org/curated/en/716141468288635541/Recovering-strong-positive-trends-in-poverty-and-opportunity

Pervaiz, A., Rahman, P., & Hasan, A. (2008) *Lessons from Karachi: The Role of Demonstration, Documentation, Mapping and Relationship Building in Advocacy for Improved Urban Municipal Services*. London, International Institute for Environment and Development.

Rana, S. (2017) 6th census findings: 207 million and counting, *The Express Tribune*, 25 August. Available at: https://tribune.com.pk/story/1490674/57-increase-pakistans-population-19-years-shows-new-census/

Sansom, K. (2011) Complementary roles? NGO-Government relations for community-based sanitation in South Asia, *Public Administration and Development*, 31(4), pp. 282–293. https://doi.org/10.1002/pad.609

Shaheen, F. (2018) Comparing informal sector engagement across Pakistan's largest urban centers: Lessons in state and non-state engagement from Karachi and Lahore, in Grant, B., Liu, C. Y., & Ye, L. (eds.), *Metropolitan Governance in Asia and the Pacific Rim: Borders, Challenges, Futures*. Singapore, Springer, pp. 95–122. https://doi.org/10.1007/978-981-13-0206-0_6

UN DESA. (2018) *World Urbanization Prospects: The 2018 Revision, Online Edition*. New York, United Nations. Available at: https://population.un.org/wup/Download/

UNDP. (2018) *2018 Human Development Index (HDI) Statistical Update*. New York, United Nations. Available at: http://hdr.undp.org/sites/default/files/2018_human_development_statistical_update.pdf

World Bank. (2013) *Pakistan – Towards an Integrated National Safety Net System*. Washington, DC, World Bank. Available at: http://documents.worldbank.org/curated/en/877951468286816490/Pakistan-Towards-an-integrated-national-safety-net-system

Yousafzai, A. (2019, April 9) Fishing communities are caught in the net of changing times. *The News International*. Available at: https://www.thenews.com.pk/print/455211-fishing-communities-are-caught-in-the-net-of-changing-times

8 Market-led development

Shakeel Ahmad[1]

Markets and the private sector could play an instrumental role in social development. This chapter argues that donors spending for developing markets for social and economic development is one of the better spending models. It presents the story of how the microfinance market in Pakistan has evolved through donor funding. It discusses how donors have aligned their support to the different phases of the development of the microfinance industry from pure grant-based funding, to limited subsidies, and to purely commercial lending. In doing so, the donors have not only developed self-sustaining microfinance institutions but have also been able to leverage limited financing for a much larger impact at scale. The chapter provides examples and case studies from Pakistan, where aid was used to help establish self-sustaining and profitable social businesses generating social impact at scale. Developing and strengthening markets for the delivery of social goods has emerged as an innovative model for the "better spending" of aid that yields high social returns for donors and recipient countries alike.

Keywords

Social development, markets, private sector, microfinance, Sustainable Development Goals, impact financing

Introduction

In December 2012, United Nations Development Programme (UNDP) undertook the impact assessment of one of its Rural Development Projects which was concluded five years back. The objective was to assess the post-project impact and sustainability of activities or models introduced by a five-year project titled Lachi Poverty Reduction Project (LPRP), implemented during 2002–2007. The main findings of the evaluation were that the microfinance model of the project had survived and was still operating on its own. The microfinance programme was up and running and was providing services to its clients even after the closure of the project five years back.

The microfinance component of LPRP was a small component: not more than 10% of the project budget was spent on it. However, it was different from the rest of the project's components in the sense that it was a loan programme, whereas all other components were based on grants. The model was such that grant money the project received was converted into a "revolving credit pool." The money was lent to the clients as a standard microfinance product. It was a self-sustaining enterprise recovering its operational and capital costs from the loan operations.

The stories of other microfinance institutions are not different from LPRP's microfinance programme. Today, the microfinance industry in Pakistan is estimated to have a potential market size of 20.5 million clients. It has reached this level over a period of approximately 22 years. As of December 2018, the microfinance industry has had a client base of 6.9 million active borrowers, gross loan portfolio of Rs. 274.7 billion (around US$2 billion),[2] and a total assets base worth of Rs. 330 billion (around US$2.35 billion). The industry may not be the biggest or most fascinating in South Asia, but it is a great example of how international aid has built a self-sustaining and self-expanding microfinance market.

In this case story, we will present how the donors' money was used to develop the markets for microfinance services. We will provide an account of the evolution of the microfinance industry and the different modes of financing that the aid money has taken at different stages of the development of the microfinance sector in Pakistan. We also provide a few successful examples of individual microfinance institutions.

The traditional model of spending aid money

Generally, the use of aid money has been in two contexts: humanitarian and development. Humanitarian contexts often result from natural or man-caused disasters. For example, donors' funding in the aftermath of earthquakes, floods, drought, food shortages, and so on. Examples of man-caused disasters include civil wars and tribal disputes or governance failures resulting in wars or famines, like the wars in Afghanistan, Syria, and Yemen – just to mention a few.

Development contexts are those where there is overall peace and security in a country, but governance failures result in inefficiencies in service delivery, economic growth, or poverty alleviation. In such cases, the donors' money is mostly used for building institutions and improving service delivery systems or addressing the issues which often emerge in developing or emerging economies. In this chapter we discuss the use of aid money in the development context.

In both development and humanitarian contexts, donors' resources take the form of grants. These grants fund development projects. In the development context, one of the challenges of this approach is that as the aid money dries up, so does the project sustainability both in terms of sustaining and upscaling the project activities. The impact of aid money is therefore limited at least in two ways. First, aid money is not sustainable as it does not regenerate itself. It is a one-time grant money. Second, since it is limited in its size or amount, it cannot cover a big number of beneficiaries to make a larger impact.

However, the good news is that the use of aid money has evolved over time from activity-based project funding to investment in building institutions and developing markets. In case of institutions building and market development, the aid money is leveraged to generate larger investments from other sources and to build institutions and systems which could operate and sustain themselves by generating resources on their own. Developing markets means developing self-sustaining organizations and enabling institutions that provide social services on the basis of cost recovery and profit-making. Such markets are similar to commercial markets in the sense that they generate financial returns for themselves; the only difference is that they aim at both profits or recovering costs and a creating social impact in terms of reducing poverty and improving social development indicators.

Such markets could be called "social impact markets," and the investment made in such markets or organizations or ventures could be seen as "social impact financing." The investments could take the forms of both donor and private investments. These are called "investments" as they are meant to be returned to the investor along with returns in excess of principal amount. In doing so, the aid money builds self-sustaining markets which could permanently provide services at a bigger scale.

Aid money for developing markets: the example of the microfinance industry in Pakistan

Donor funding has been instrumental in the development and growth of the microfinance industry of Pakistan. It has facilitated the nurturing of many new entities that, with the passage of time, have become self-sufficient and profitable social enterprises. The establishment of microfinance industry in Pakistan can be traced back to late 1990s. It started off with standalone projects of poverty alleviation and gradually shaped up into a formal sector with the establishment of the first specialized microfinance organization, called the Kashf Foundation, in 1996. With a grant from Grameen Bank, the Kashf Foundation set out to alleviate poverty through women-centred initiatives that provided them with access to formal financial services and skills-development support. Since its establishment in 1996, the Kashf Foundation has expanded its coverage to all four provinces of Pakistan through a network of 260 branches. To date, it has reached out to 1.5 million households and disbursed over PKR 66 billion through 3.3 million loans to people at the bottom of the income distribution chain.

Similar to Grameen Bank, the Kashf Foundation uses the model of peer groups for insuring lenders against default. Focusing largely on women, it does not seek any physical collateral as a prerequisite for obtaining loans but instead requires each member of the group to take responsibility for repayment of the loan on behalf of the borrower, in case of default. This model has been successful in providing loans of small value at the grass-roots level. The Kashf Foundation offers a range of products catering to clients' needs: general loans, emergency loans, and business loans. The success of Kashf Foundation lies in its ability to turn an organization into a financially sustainable business model.

Following the development of Kashf Foundation and a number of other microfinance organizations, an informal association of retailers in the microfinance industry was established in 1997 as the Microfinance Network of Pakistan. Its objective was to connect market players to exchange ideas and best practices.

The association gradually took a more formal role and developed into an entity registered with the Securities and Exchange Commission of Pakistan. It now has a membership of 46 retail microfinance providers – evidence of the growth and maturity gained by the industry.

In the initial stages, the Pakistan Microfinance Network (PMN) received financial support from Aga Khan Foundation and the Asia Foundation. However, more recently, it has received and continues to receive support from the UK Department for International Development (Pakistan Microfinance Network, 2017a; Devex, 2018). Through DFID support it acts as a channel for strengthening the capacity of its members while also improving the regulatory environment to allow for the expansion of the microfinance industry. The PMN serves as a facilitator that connects consumers to microfinance institutions, and acts as an information hub and as an advocate for microfinance, supporting the growth and maturity of the sector.

In 2000, the government also undertook efforts to formalize the industry by developing an apex body called the Pakistan Poverty Alleviation Fund (PPAF) to act as an intermediary with a role of a wholesaler. To date, it supports retailers in achieving access to finance while building their capacity for extending funds to consumers of microfinance. While PPAF was primarily funded by the World Bank, in the same year, the government also established Khushali Bank through financial support from Asian Development Bank. When the State Bank of Pakistan passed the Microfinance Institutions Ordinance 2001, it formally regularized the market and provided less stringent requirements for the entrance of new private microfinance players. This was a major achievement for the industry as it evolved from a purely social enterprise to a commercial business (Akhuwat, 2018; Pakistan Microfinance Network, 2008).

Regarding microfinance banks, Khushali was a government-owned bank. Its development was soon followed by the first private sector bank, called the First Microfinance Bank Ltd, in 2002; it was a transformation of Aga Khan Rural Support Programme's (AKRSP) credit and savings business. The First Microfinance Bank uses a variety of credit models to address the needs of its clients: individual lending, group-based lending, and village banking. On the other hand, Khushali Bank follows only the Grameen Bank's peer group model to issue credit. It engages NGOs to develop peer groups at the grass-roots level, which are supported by the bank's loan officers.

These different lending models have strengthened the microfinance sector in Pakistan and now offer a space for innovation. In 2001, Akhuwat, a microfinance organization using an Islamic mode of financing, was established. Akhuwat operates a completely different model: it creates capital for lending and operations through charities and philanthropic contributions. This model may not be a direct product of donors' money, but it has undoubtedly emerged out of the vibrant microfinance industry supported primarily through aid. Akhuwat uses religious institutions such as mosques and churches as the centres for interacting with clients and engaging in loan disbursement. It facilitates access and development in areas of education, health, population welfare, housing, and employment generation. Today, Akhuwat is one of the leading organizations in the industry with the highest number of active borrowers recorded in 2017 (820,000 borrowers). Its institutional strength and network have also been leveraged by governments to extend credit at the grass-roots level through public funds.

The microfinance industry is playing a leading role in facilitating financial inclusion of a huge proportion of the unbanked population in Pakistan. The expansion in the industry, facilitated through donor funding, can be seen from the indicators in Table 8.1: Growth of the microfinance industry. The number of active borrowers has increased from 1.6 million in 2010 to 5.8 million in 2017, and correspondingly the average gross loan portfolio for the industry also increased from 18.6 billion in 2010 to 164.1 billion in 2017. It's interesting to note that active women borrowers have also consistently increased, going from being 0.8 million in 2010 to 2.7 million in 2017. This trend demonstrates the potential of microfinancing to facilitate women empowerment.

Financial inclusion has also improved in Pakistan through donor support. In 2017, the World Bank started a Financial Inclusion and Infrastructure Project in Pakistan with the primary objective to increase financial inclusion and support State Bank of Pakistan (SBP) in achieving its target of increasing financial access to at least 50% of the adult population by 2020. Under the project, US$50 million dollars were allocated to establish a line of credit for microfinance banks and non-bank microfinance institutions to ease liquidity constraints and use the fund as an incentive to attract private investment.

Under the same project, US$10 million was also set aside to increase access to finance small- and medium-sized enterprises. This resource pool was set aside to support a DFID-funded project under which a credit guarantee scheme was provided for small and rural enterprises. Operated by SBP, the credit guarantee scheme was covering up to 60% of the loan value that was extended to small, micro, and rural enterprises, and to especially those operated by women. Effectively, aid was being used to encourage extension of high-risk microcredit by guaranteeing against a major share of default. The terms and conditions of the guarantee were further modified by providing a higher guarantee on loans with low collateral to encourage inclusion of high-risk borrowers (State Bank of Pakistan, 2017; The World Bank, 2017).

Table 8.1 Growth of the microfinance industry

	2010	2011	2012	2013	2014	2015	2016	2017
Active borrowers (in millions)	1.6	1.7	2.0	2.4	2.8	3.6	4.2	5.8
Average gross loan portfolio (PKR billions)	18.6	22.7	29.8	40.4	56.0	76.8	110.7	164.1
Active women borrowers (in millions)	0.8	0.9	1.3	1.4	1.6	2	2.3	2.7
Branches	1,405	1,550	1,630	1,606	1,747	2,754	2,367	3,533
Total assets (PKR billions)	35.8	48.6	61.9	81.5	100.7	145.1	225.3	330.4
Deposits (PKR billions)	10.1	13.9	20.8	32.9	42.7	60	118.1	185.9

Source: Pakistan Microfinance Network, 2017b

Similar projects should be promoted where aid can be used to facilitate development of livelihood sources of the poor and marginalized groups through activities that target grass-roots-level interventions. As these projects are operated through public and private institutions, they also help in strengthening institutional capacity and quality of governance. Taking a cue from the examples discussed, an independent evaluation commissioned by DFID in the credit guarantee scheme for SMEs helped provide a good assessment of the factors that were hindering higher uptake of the credit scheme and corrective action was taken accordingly.

The approach to the development of Pakistan microfinance industry

The development of the microfinance industry in Pakistan provides a classic example of the use of donor funds as "leverage points" and for building markets. In 1999, Pakistan's microfinance sector was serving only 60,000 active borrowers. In 1996, the UK Department for International Development (DFID) provided seed money to the Kashf Foundation, a microfinance institute based out of Lahore in Punjab province of Pakistan. Kashf simply replicated the Grameen Bank model of group lending focusing on women. After a successful two-year pilot phase, DFID provided a grant of £3.2 million over a five-year project which was used as a "revolving credit pool" and for funding operational costs. The donor approach to building the microfinance industry could be divided into three phases as elaborated below.

Three phases of aid investments

First phase (roughly 1980s–2002): grants for funding whole operations (credit plus operational cost)

The microfinance market has evolved over a period of time with donors' lines of services and support varying during each phase. The first phase is the one where the donors provided grants for building a "revolving credit fund" and meeting operational expenses. The revolving fund took different forms. In South Asia Poverty Alleviation Programme (SAPAP), which was a predecessor of LPRP, UNDP provided a grant facility for establishing a "village-based credit revolving fund." The amount, though small in size, was provided to a village organization in Shakardara, Kohat district of Khyber Pakhtunkhwa province of Pakistan. A small committee of the members of the community organization was trained on bookkeeping and accounting to manage the credit fund. In this model, the members of the village organization could get a maximum loan of Rs. 10,000 (US$150 at that time) for a maximum period of 12 months, to be repaid in instalments consisting of principal plus interest payments. A village activist served as a credit officer, recovering loan instalments and making deposits in the bank account maintained in the name of the village organization. A professional staff from the project was used to provide technical support to the village organizations for credit operations.

A similar model was used at a larger scale. Bilateral donors like DFID provided similar support to a number of organizations in Pakistan. For example, LPRP, which was funded by DFID and managed by UNDP, provided a credit line for microfinance operations besides the provision of grants for other components of LPRP like community-based infrastructure, skills training, and agriculture. Grants were provided for setting up a revolving credit fund at the project level, which was later on handed over to a local NGO named Sarhad Rural Support Programme.

The project extended loans to the members of the grass-roots community organizations. The amount (principal plus interest) recovered from the clients made up the "revolving credit fund." The idea was that the fund would be used for credit operations in a sustainable manner where the organization/microfinance institution (MFI) would recover its operational and other costs through interest. In this way, the credit operations would continue and expand on its own. A number of today's big organizations emerged from this form of donors' support, including the Kashf Foundation.

Even before these later-stage initiatives, the foundation for this model was laid by Aga Khan Rural Support Programme (AKRSP) and Orangi Pilot Project in 1980s. The AKRSP, established in 1982 by Agha Khan Foundation, was one of the first NGOs in Pakistan to use the model of village-based lending, and it inspired the development of other Rural Support Programmes. With a focus on microcredit, AKRSP developed a number of village organizations to upscale its poverty alleviation work through an integrated approach that worked across education and health sectors, and employment-generating initiatives. It was one of the first successful models that was able to replicate the model of "revolving funds for credit" across a number of villages.

On the other hand, Orangi Pilot Project (OPP) aimed at addressing urban poverty in Karachi, the metropolitan city of Pakistan. Established in 1987, Orangi Pilot Project extended credit in urban slums for affordable housing, education, and sanitation. Given the specialized nature of development through microcredit, Orangi Pilot Project led to the development of a specialized subsidiary called Orangi Charitable Trust in 1989, with an initial loan from National Bank of Pakistan. The trust used the model of individual lending for its client base as opposed to the group-based lending models which were prevalent during this era in Pakistan and outside. Orangi Pilot Project is also considered one of the innovative models in addressing urban poverty.

Second phase (2000–2015): subsidized credit lines

In this phase, the donors mostly supported "apex organizations" to provide "wholesale" lending to MFIs. The most significant example is the Pakistan Poverty Alleviation Fund (PPAF) established in the year 2000. PPAF, established by the government of Pakistan as non-for-profit company, received the larger component of its funding from the World Bank. The World Bank lent money to the government in the form of soft loans and the government forwarded it to PPAF, which used a portion of it as a "revolving credit fund" for onward lending to the MFIs. PPAF provided credit lines to the MFIs along with grant money (for other activities) proportionate to the loan amount and to incentivize loan operations.

The credit line was wholesale lending at a fixed interest rate. The MFIs further lend this money at an increased interest rate to cover its operational cost, risk, and capitalization requirements. At this same time, the grant money was used for other community-based interventions like community-based infrastructure development, agriculture development, enterprise development training, and so on. The grant money was therefore used to supplement credit operations and subsidize the operational costs related to it. Additionally, it was used as a leverage point to generate demand for credit.

This model had a number of benefits. First, it was the first step in commercializing the credit programmes. Until, now the credit support to the MFIs was completely grant-based and thus 100% subsidized. It therefore set the basis for market-based lending models. Second, it was a step in making the credit operations self-sustaining at both the "retail level" of the MFIs and at the "wholesale level" of the apex organization. At the retail level, the MFIs were forced to think of credit as a banking operation to recover both principal amount and interest in a way to meet both its capital needs and operational costs. Those organizations which received such credit lines were encouraged to separate their lending operations from their grants-based operations. For the apex body, the benefit was again in the form of building a self-sustaining credit pool which could serve a larger clientele base both in terms of "service providers like MFIs" and "grass-roots consumers/poor population."

A number of commercial banks in Pakistan also entered into the microfinance industry at this stage to experiment with the "wholesale" credit line extension model. This did not sustain for a long time, but there are a few situations worth mentioning. Habib Bank was one of the first banks to undertake this initiative of indirect lending. It extended a credit line worth of Rs. 2.2 billion to National Rural Support Programme for on-lending to rural communities in the form of small loans between 1999 to 2000. The Bank of Khyber also used this model to provide wholesale funds to NGOs for grass-roots-level interventions by mobilizing funds from a number of donors. For instance, it extended a credit line of Rs. 10 million to SRSP for on-lending. Eventually, however, the emergence of PPAF somewhat reduced the role of commercial banks acting as "wholesalers" of microcredit.

This era sets the foundation for an entire commercial microfinance industry. During this period and afterwards, the microfinance operations took an entirely different tangent from a "recovering cost model" to a "making profit model." During this period, the institutions both at the retail and wholesale levels were using operational models with the view to recover credit operational costs. In the next era, the microfinance industry evolved into a commercial sector for profit-making.

Third phase (2016 onwards): commercial or for-profit lending

In this phase, the microfinance industry evolved into a commercial lending sector. It was a period (this period also overlaps with the above phase) where the State

Bank of Pakistan (SBP) introduced legislation for the MFIs and other entities to register as microfinance banks (MFB). The MFB model allowed the microfinance organizations to diversify their products, and they were allowed to take deposits/savings from the consumers just like conventional banks. This benefited the MFBs in at least two ways. First, it increased their sources of funding. Earlier, the MFIs used to mobilize credit lines only from donors or banks. Now they could also mobilize funds themselves at the local level. Second, it also helped them in increasing the sources of their returns. They could invest savings/deposits in other instruments of capital markets like bonds and equity markets to generate profits for themselves.

During this period, the donors' support focused on the development of MFB industry through different means. In some cases, the donors provided technical and financial support to the SBP for policymaking and for creating an enabling environment for the development of the MFB industry. For example, the World Bank supported a Financial Inclusion and Infrastructure Project in 2017 with the primary objective to increase financial inclusion in Pakistan and support State Bank of Pakistan (SBP) in achieving its target of increasing financial access to at least 50% of the adult population by 2020. Under the project, US$50 million dollars were allocated to establish a "credit line" for microfinance banks and non-bank microfinance institutions to ease liquidity constraints and use the fund as an incentive to attract private investment. Both DFID and the World Bank (WB) used the project platform to convene discussions with key stakeholders to identify areas for further improvement and supported SBP in strengthening the overall policy and regulatory environment for the growth of the microfinance industry.

In other cases, the donors helped the SBP in developing innovative risk mitigation instruments for the MFBs. For example, the donors helped create the "credit guarantee line" at the SBP, which was used to "de-risk" the credit operation of MFBs. This facility was to incentivize the MFBs in providing mostly non-collateralized loans to marginalized geographical areas and segments of populations generally considered risky for loan recovery. The amount was used to cover the losses of MFBs from such credit operations. Under the same WB supported Financial Inclusion and Infrastructure Project, US$10 million was also set aside to increase access to finances for small- and medium-sized enterprises. This resource pool was set aside to support an ongoing DFID-funded project under which a credit guarantee scheme was provided for small and rural enterprises. Operated by SBP, the credit guarantee scheme covered up to 60% of loan value that was extended to small, micro, and rural enterprises, and to especially those operated by women. Effectively, aid was being used to encourage extension of high-risk microcredit by guaranteeing against a major share of default. The terms and conditions of the guarantee were further modified by providing a higher guarantee on loans with low collateral to encourage inclusion of high-risk borrowers (State Bank of Pakistan, 2017; The World Bank, 2017).

During this phase, the donors also made available resources for "research and development" for innovation in products and microfinance delivery models. For instance, Karandaaz, a DFID- and Bill & Melinda Gates Foundation–funded for-profit organization, is working on a number of initiatives to advance techno-logical solutions for financial inclusion. It not only undertakes vital research on different sectors in need of financial support but also organized fintech challenges to bring out innovative solutions that can be piloted through donor funding.

Another hallmark of this phase is the entry of some of non-conventional play-ers into the MF market – for example, the Easypaisa service by Telenor, which quickly became a pioneer of branchless banking in Pakistan. The Easypaisa ser-vice allows users to have a mobile account which can be used to transfer money to any other Easypaisa account or to a Computerized National Identify Card number or any bank account across Pakistan. It leverages its network of more than 70,000 shops to allow its users to easily send or receive money through their local retailer. Its mobile app now allows a variety of services: making online bill payments, giv-ing donations, purchasing cinema or bus tickets at a discount, and using savings and insurance services.

Easypaisa is an example of a digital innovation in branchless banking which has helped in reaching out to a large unbanked population and is quickly become a one-stop digital platforms for all types of transactions. It not only saves costs for the service providers but also increases the outreach to MF services in far-flung areas. As of October 2018, this product had reached 1 million active users and recorded total transactions worth Rs. 1.2 trillion in 2017. Overall, this period has shown a lot of innovation in the MF industry.

The most recent development in this phase has been the donors' investments taking the shape of "equity." Until now, donor support has been in the form of "credit lines" and "technical support" for creating an enabling policy environment for the MF industry. Now, donors have started equity investment in "for-profit" entities. Let's discuss two examples here: one is the Pakistan Microfinance Invest-ment Company and the other is Karandaaz.

The Pakistan Microfinance Investment Company (PMIC) was established through a joint partnership of Pakistan Poverty Alleviation Fund, Karandaaz Paki-stan, and KfW Development Bank to support the implementation of the National Financial Inclusion Strategy. All three entities are shareholders of the organization with PPAF being the major shareholder, owning 49% of shares, and Karandaaz and KfW bank owning 37.8% and 13.2% of the shares, respectively. The PMIC works as a national apex institution for microfinance institutions in Pakistan. It has a for-profit structure in which it mobilizes funds from local and international capi-tal markets and advances market-based financing solutions to local microfinance organizations of Pakistan. It also provides technical assistance to microfinance insti-tutions in strengthening values chains, developing new products, and leveraging technology for cost-efficient solutions and higher impact.

Karandaaz, is an organization registered under Section 42 of the Compa-nies Act. It was established in 2014 through funding from DFID and the Bill & Melinda Gates Foundation to increase access to finances for micro-, small-, and medium-sized enterprises through commercial markets and support the use of technology and digital platforms to increase financial inclusion. It works as a for-profit organization across four main streams: capital provision, digital platforms, knowledge management and communication, and innovation. Through its stream of Karandaaz Capital it acts as a direct investor in Small and Medium Enterprises (SMEs), as a wholesaler to reach out to sector specific clients, and as a strategic investor to promote development of financial markets for SMEs. PMIC is an example of Karandaaz initiatives through which it strengthens markets for easy access to finances. Karandaaz has also funded research and innovation in multiple digital platforms that can ease access to finances and digitize payment solutions.

Advantages of developing markets

The net official development assistance (ODA) received by Pakistan in 2017 was only 0.17% of GNI. In per capita terms, this equates to only US$12. The amount of ODA received by Pakistan has declined from 10.4% of GNI in 1963 to a meagre 0.17%. Simultaneously, the population has increased massively: from 42.9 million in 1960 to 207.7 million in 2017. The decrease in aid coupled with extremely high rise in population means that aid is very thinly distributed across the population.

The mechanisms of aid transfer also make it more challenging for donors to ensure that the highest share of aid is received by the large masses of the poor population. Consequently, aid fails to generate impact at scale as a result of being allocated to projects that can only be sustained for as long as foreign aid is being supplied for them. With sustainability being the biggest challenge, profit-generating markets are the best option to ensure continuity and expansion beyond aid. In the latest 2017 report of Asian Development Bank on development effectiveness, the bank estimated that only 66% of its completed sovereign projects can be rated as being sustainable. The report also states that sustainability ratings are the highest in industry and trade, thereby strengthening the argument that the private sector is the best investment of foreign aid.

A relatively small foreign aid investment in for-profit social enterprises can be leveraged for yielding high and continuous returns, as seen in the examples of microfinance institutions. To put this into perspective, the recent purchase of 82.8% of the shareholder equity of Kashf Microfinance Bank by FINCA International was valued at US$8.4 million (PKR 824.7 million), showing the approximate net worth of the entity. The microfinance institutions in Pakistan generate revenues sufficient to cover their expenses, costs of capitalization, and other expenses. The latest annual report by Pakistan Microfinance Network shows that on average operational self-sufficiency is at 124.7% and financial self-sufficiency is at 122.4%, highlighting the ability of microfinance institutions to achieve financial self-sustainability. The MFIs on average generate almost 77% of the revenues from a loan portfolio and 17% from other financial services.

Markets are also a good choice for investment as they promote competition, efficiency, and innovation. Innovation in social products and service delivery not only benefit the masses but also create economic incentives for investors to keep expanding. For instance, the Pakistan Microfinance Network continuously explores ways to improve service delivery and the regulatory environment for microfinance industry. In one of its recent developments, the network provided support in the digitization of the entire end-to-end system of transactions used by microfinance institutions. This will help in strengthening credit scoring of consumers and through improved programming, help microfinance institutions in increasing outreach and extending loans with more confidence in recovery prospects (Pakistan Microfinance Network, 2017a, b).

It is well-recognized that expansion of the microfinance industry has supported the growth of the small- and medium-sized enterprises in Pakistan. The new private players are using innovative tools to extend outreach, facilitate transactions, and reduce operational costs; rapid expansion of branchless banking (mobile wallets) and digital credit systems, including the transfer of nano-loans through mobile phones, are examples of technological advancements being pushed through the industry.

The Telenor Bank, previously known as Tameer Bank, was the first microfinance bank of Pakistan to launch a mobile banking service, called Easypaisa, in 2009. Since its launch, the users of Easypaisa have increased rapidly as it continues its outreach to a huge proportion of the low-income population that was previously unbanked. The Telenor Bank has recorded a total of PKR 1.2 trillion transactions through its Easypaisa service with a network of 117,424 agents spread around the country. With a mobile app attached to it, the service has helped expand e-commerce in the country and public service delivery, providing a range of easy transaction possibilities to its customers. These examples of innovation demonstrate how aid can generate a multiplier effect in product development leading to a host of new services and business ideas across the value chain.

Market constraints

There is no doubt that the markets play critical role in social development. There are, however, sensitivities which need to be kept in mind. Businesses are there for profit-making. They may use donors' money to maximize their profits without any social impact. It is quite difficult to evaluate the on-ground impact of public money being invested in the private sector. As donor money flows through equity funds or financial institutions before being provided to the private sector in the form of grants or loans, it becomes increasingly difficult to measure value for money in the form of grass-roots investment in social development. To avoid this issue, aid money should be linked to tangible outcomes. Impact measurement should be an integral part of the overall business operations.

Second, this mechanism of aid transfer also blurs the lines between a *profitability* motive and giving aid with the objective to eliminating poverty. This is especially pertinent in the case of providing aid to the private sector for delivering public services such as health and education. Critics argue that such services should remain detached with the objective of profitability. To address these concerns, Eurodad has developed some guidelines of which the important features are as follows (cited in The Reality of Aid, 2012):

* Private sector investment should be aligned to the priorities of the developing countries.
* The development outcomes should be the priority for selection and evaluation, while complying with responsible investment standards.
* Improve the transparency standards of financial intermediaries and evaluate the use of funds through strict standards.

From the perspective of aid recipient countries, the model for direct investment in private businesses can only result in high returns if partners engaged do not have any vested interests or engage in rent-seeking behaviour. This is an important element for private financing to have a high social and economic return. In Pakistan some incidents of corruption in use of aid have surfaced that have discouraged donors from financing socially motivated private or public enterprises. An unfortunate incident surfaced when a judicial inquiry into UK funded education efforts in Sindh revealed how aid was being channelled for 5,000 schools and 40,000 teachers that only existed on paper. This challenge was not only seen in public education initiatives but also in private businesses where the teachers were paid less than one-fourth of minimum wage without any form of transparency.

Last, while a strong business case is available for using aid to develop private institutions, it is also important to assess the effects of aid money on free market forces. Aid transfer can distort market forces and lead to crowding out of private investment. Hence, incentives for aid-receiving individuals should be aligned to productive activities to discourage rent-seeking behaviour. Large inflows of aid into private businesses can mask inefficiencies and weaken the objective to develop sustainable and inclusive businesses.

Conclusion

Ultimately, the end goal of donor funding is to eradicate poverty and achieve sustainable development. The private sector and markets are the driving forces needed to achieve this objective as they hold the potential to create a larger-scale impact in a self-sustaining manner. Realizing its central position to the paradigm of sustainable development and the achievement of UN Sustainable Development Goals (SDGs), the donor countries are increasingly partnering with the private sector. The donor governments and organizations are now increasingly investing in markets through channels that support non-public entities that have social development as their core business objective.

As seen in the many examples quoted in this chapter, Pakistan has experienced the establishment and maturity of microfinance market through donor funding for markets and private sector development. Today, a dynamic microfinance industry exists in Pakistan and donors have made significant contributions to this achievement. At the start of the process, aid money was used as grants for setting up "revolving credit pools" and developing microfinance organizations. It was followed by using aid money as repayable credit lines to microfinance organizations.

Since the microfinance industry has reached a level of maturity, the aid money is now being used as investments in commercial and for-profit ventures. This has all led to the emergence of a self-sustaining and self-expanding market for microfinance. The market of the microfinance industry is the best example of "better spending" of donors' money for social development and poverty eradication.

Notes

1 The author would like to acknowledge the contribution of Aroub Farooq in compiling this chapter. Aroub has master's degree in international economic policy and analysis from the University of Westminster, UK. She works on development issues, ranging from multidimensional poverty, inequality, economic growth, urbanization, and implementation of Sustainable Development Goals. She is currently employed as a research analyst at the United Nations Development Programme, Pakistan.
2 The exchange rate of Rs. 140 per dollar was used.

References

Akhuwat. (2018) State of microfinance in Pakistan. Available at: https://www.akhuwat. org.pk/state-of-microfinance-in-pakistan/ [Accessed 30 March 2019].

Devex. (2018) Pakistan microfinance network. Available at: https://www.devex.com/organizations/pakistan-microfinance-network-pmn-73953 [Accessed 25 January 2019].

Pakistan Microfinance Network. (2008) MicroNote No. 2. Regulated playground for branchless banking in Pakistan. Authored by Sara Saeed Khan. Available at: http://www. microfinanceconnect.info/assets/articles/MicroNOTE%202%20-%20Regulated%20 Playground%20for%20Branchless%20Banking%20in%20Pakistan.pdf [Accessed 15 August 2019].

Pakistan Microfinance Network. (2017a) Annual report 2017: Achieving together. Available at: http://www.microfinanceconnect.info/assets/articles/6b51d70c468bf7a7f984a1 94d0a6ee26.pdf [Accessed 25 January 2019].

Pakistan Microfinance Network. (2017b) Pakistan microfinance review 2017: Annual assessment of the microfinance industry. Available at: http://www.microfinanceconnect. info/assets/articles/da6af2f0f19ca541bbc04db2f158cf98.pdf [Accessed 30 March 2019].

State Bank of Pakistan. (2017) Annual report fiscal year 2017. Available at: http://www. sbp.org.pk/reports/annual/arFY17/Vol-1/Chapter-4.pdf [Accessed 09 April 2019].

The Reality of Aid. (2012) *Aid and the Private Sector: Catalysing Poverty Reduction and Development*. Philippines, Reality of Aid Network Coordination Committee.

The World Bank. (2017) Project information document/Integrated safeguards datasheet: Pakistan Financial Inclusion and Infrastructure Project. Available at: http://projects. worldbank.org/P159428/?lang=en&tab=documents&subTab=projectDocuments [Accessed 9 April 2019].

Part IV
Resilient communities in fragile states

Central South Asia

Part IV

9 Community-driven development as a mechanism for realizing global development goals

The National Solidarity Programme and Citizens' Charter Afghanistan Program

Nipa Banerjee

Afghanistan, a country emerging out of a long civil war, followed by the autocratic extremist Taliban rule was in need of good governance and community-building in 2001. The National Solidarity Programme (NSP) was conceived to give Afghan villages a democratic local administration, along with access to basic services. In 2003, NSP was launched as a programme of the Afghan government to reduce poverty through establishing and strengthening a national network of effective self-governing community institutions – Community Development Councils for local governance and socio-economic development. NSP developed the ability of rural communities to identify, plan, manage, monitor, and maintain their own development projects, making the development results sustainable. NSP and the successor programme – Citizens' Charter Afghanistan Program (CCAP) – empower the rural communities to make decisions affecting their own lives and livelihoods, and they collectively promote the global development goals through local actions. The process is inclusive, leaving nobody behind. Here goes the story of NSP with its roots in distant Indonesia, its journey to Afghanistan, and its continuing trajectory under CCAP.

Keywords

Community-driven development, Millennium and Sustainable Development Goals, nationalization of global goals, end of poverty, cost-effective spending, value for money

State legitimacy-building through public service delivery in solidarity with the people

The National Solidarity Programme (NSP) is a surprise development success story that emerged out of Afghanistan, one of world's poorest countries and the deadliest for terrorism. The foundation of NSP – a community-driven development programme – was laid to spur development in war-torn Afghanistan, after the fall of the Taliban. In the immediate post-conflict period, the goal of the transitional government of Afghanistan was to build a state which through delivery of essential services to its citizens, would earn people's support and gain legitimacy in the eyes of the citizens. The legitimacy earned would lend the state the power to govern with authority.

One of the priorities in Afghanistan's state-building agenda in 2001 was, thus, to extend the reach of the government services across the country to improve people's lives, a spirit underlying the global development goals. But to be successful, such a development undertaking in Afghanistan would require strong institutions and tools, essentially lacking in the country at that point. Decades of war having largely destroyed institutions in Afghanistan, one of the first essentials in 2001 was building institutions for delivering security, justice, and development services. For delivering development services, the chosen focus was community building, along with fostering the spirit of community participation in the local development process.

A participatory process would develop the ability of the local communities to identify community-level development needs, plan the ways to address these through community dialogues, and use their own and Afghan government resources to implement the planned projects. The community members, who would propose and plan the projects, would be in the best position to manage and monitor the development activities. Such a process, empowering the communities to make decisions affecting their own lives and livelihoods, would foster local accountability and ownership of the community development process. The process would help build social capital and strengthen local governance

Building blocks for realizing an Afghan dream

Building a legitimate state, in solidarity with the Afghan communities, was the dream of Afghanistan transitional government's powerful finance minister, Ashraf Ghani, who became the president of the country in 2014. His priority was to deliver government services to communities across the country, essentially with the involvement and participation of the communities themselves.

An American-educated and financially well-off employee of the World Bank, Ashraf Ghani had returned to Afghanistan in 2001 to support his homeland's pursuits in building an effective state to better serve the people, who in turn would support the government. Upon a review of the global experiments with governance practices to better meet citizens' needs, Ghani decided to call his long-time friend and World Bank colleague, an American expert in international development, Scott Guggenheim, to help build the base for a state that would be connected to the people, who, in solidarity with the government, would help build a new Afghan nation.

At the time, Guggenheim, basically an expert on Indonesia, knew little about Afghanistan other than what he learned from Rudyard Kipling's narratives from 1888. Keeping in mind the central importance of delivery of essential services to the citizens for state building, he drew on his past experiments with programmes for people's benefits and social capital development. He zeroed in on his experience with the Kecamatan Community Development Program (KDP). Afghanistan's most successful development programme – NSP – thus, has its roots in distant Indonesia (Guggenheim et al., 2004).

Guggenheim, in collaboration with the World Bank, had founded KDP in Indonesia in 1998 in the aftermath of the Asian Financial Crisis. KDP was not initially planned in response to conflict, but since its operations coincided with the violence that followed the fall of Suharto's New Order government, KDP served as a tool for post-conflict reconstruction and development and state stabilization, following the principles of community-driven development (CDD) programmes. The CDD approach, by giving communities the power to plan, manage, and control programme budgets, promotes better spending, more efficient and effective use of resources. The participatory approach has the added benefits of developing local leadership and institutions, generating civic capacity, and developing social capital, much needed in a poor country, struggling to emerge out of a state of conflict.

Ashraf Ghani and Scott Guggenheim decided to utilize the KDP model to fight poverty in post-conflict Afghanistan using a community-driven approach that would deliver services aligned with the priority needs of the poor rural communities. Participation of community members in identification of community needs, and in the design and implementation of programmes to address these, would help attain this goal. Afghanistan's population being overwhelmingly rural and poor, with small agriculture-based livelihoods but lacking supplies of water for farming and drinking and other basic amenities, needed immediate attention. Village-based organizations would be best able to address these inadequacies.

The start-up

The National Solidarity Programme (NSP) was started in 2003 by the government of Afghanistan, under the auspices of the Ministry of Rural Rehabilitation and Development (MRRD), led by one of the most charismatic Afghan leaders, Mohammad Haneef Atmar, who later served as an advisor with the Ministries of Education and Interior and as the National Security. NSP thrived under his leadership and guidance.

The goal of NSP was to strengthen the Afghan government's state machinery for service delivery to the rural population. NSP was created under the authorization of an article of the Afghanistan constitution that underlines the need for the state to undertake programmes for industrial development, economic growth, and betterment of the living standards of the people (Sadri, 2019). The NSP mandate was to establish community development organizations in rural settlements to identify people's needs and plan and implement programmes to address these. A national network of such organizations, named Community Development Councils (CDCs), was to be established to empower communities to deliver critical services in the rural areas. These CDCs, in consultation with the community members, would identify projects and secure government funding for implementation of the projects to improve communities' access to infrastructure, markets, and basic services.

The foundation of the NSP was the *community*. NSP grouped some 40,000 rural settlements in the country into 28,500 communities, each of which was to establish a CDC. A rural settlement of at least 25 households and up to a maximum of 300 would constitute a community. These communities were organized around the CDCs, each of which was a body elected by community members by secret ballot. These CDCs were at the centre of NSP operations. The democratically elected council members assumed the role of an institution for local governance and socio-economic development of the village. The CDCs, people's organizations, served as the government's arm for service delivery locally.

The National Solidarity Programme, global development goals, and the end of poverty

It has been established that the NSP approach offered a low-cost solution to lack of basic services in remote areas, and thereby, made contributions to poverty reduction, the principal tenet of the Millennium Development Goals (MDGs) and the Sustainable Development Goals (SDGs), the 2030 Agenda. It is interesting that although the Afghan government had signed the Millennium Declaration for MDG implementation, it is unclear if at the conception stage of NSP at the start of the millennium, Afghan planners had consciously integrated the MDGs into the planning process of NSP. A conscious process of what Afghans name *nationalization* of SDGs started nearly a decade-and-a-half later. But the spirit of poverty reduction through provision of essential services to meet the basic needs of the poor was central to NSP from the very beginning. The plan was to address the various dimensions of poverty that negatively affect people's lives and well-being: poor health, lack of access to education, lack of income, and so on, among many other needs and wants that the global development goals (MDGs and SDGs) address.

A people's participation-oriented process gave CDCs strong local roots, helped generate local ownership and accountability, and promoted potentials of sustainability of the results of the projects implemented. The best indicator of this claim lies in the unhampered operation of the CDCs under NSP for a decade-and-a-half and their continued operation, with an expanded mandate, under the Afghanistan National Priority Program, Citizens' Charter. NSP has been replaced by the Citizens' Charter Afghanistan Program (CCAP).

The process

The five phases of the NSP cycle reflect a high degree of citizen participation from project selection, implementation, and fund usage to progress monitoring and fulfilling accountability requirements (Centre for Public Impact, 2016).

In phase one, an NGO (called a facilitating partner), contracted by the MRRD of the government of Afghanistan, would be assigned to a village to mobilize and motivate the community to form a community organization. In phase two, a community organization, CDC, would be formed by each village community through an election process. Each CDC would be registered with the state. In phase three, in consultation with the community, the council members would identify local needs and establish a project list (a Community Development Plan, or CDP) to address these. Each CDC would submit its plan to NSP for funding. In phase four, the government would disburse block grants to the CDCs from the NSP budget to facilitate implementation of the community projects. The grants would be transferred to the community bank accounts. Upon receipt of the block funds, the CDCs would launch the implementation phase. They would manage implementation of the CDP, monitor projects, and report to their communities on progress of projects and use of funds. Under phase five, the facilitating partners (a group of 8 national and 21 international NGOs) would assess the technical quality of completed projects and document the lessons learned (Beath et al., 2015). One partner per CDC would be designated for this task. The facilitating partners would also undertake a series of activities to build capacities of the CDC members in various skills, such as technical skills for project implementation and management, and accounting to ensure quality project delivery and to meet accountability needs. MRRD would oversee NSP activities and ensure the government's presence in villages across the country, connecting the villages with the Afghan government. The CDCs would also be connected with aid agencies that offer funds and services to rural areas to help diversify the funding sources. The World Bank financed the NSP with contributions from major bilateral donors. The five phases normally would cover a period of three years in each village.

NSP completed three funding rounds. The first round covered the period 2003–2007 (NSP I), with the second round operating in the period 2007–2011 (NSP II). With the completion of the third round of funding, NSP transitioned to the Citizens' Charter Afghanistan Program (CCAP), which was launched in September 2016. CCAP, currently under implementation, aims to improve the delivery of core infrastructure and social services to communities through strengthened CDCs. CCAP tasks are very similar to those of the NSP I, II, and III. NSP performed under extremely difficult situation in the country, in the years following the Taliban era. While CCAP will build on the achievements of NSP, it is also expected to address certain unexplored elements which NSP could not touch. Thus, CCAP has a broader mandate.

What was gained

NSP implementation took off fast, judged both by the number of CDCs established, number of projects funded, and amount of community block grants disbursed. By 2005, US$100 million was disbursed as block grants. Each CDC could receive a grant of up to US$60,000, conditional upon a 10% contribution (in cash, labour, or material) from the community. The average size of block grants released was approximately US$35,000 dollars. By 2009, nearly 22,000 CDCs were established. From the birth of the NSP in 2003 to the year 2013, over

US$1 billion dollars was disbursed as block grants for 64,000 projects. NSP records of 2016 show further expansion of CDCs.

By 2016, slightly more than 35,000 villages (nearly 80% coverage of villages in the country) had elected CDCs. CDCs were operational in all provinces of Afghanistan and covered 98% of the country's 398 districts. Total number of males elected to the CDCs was 300,000. Just over 150,000 female members were elected. Therefore, 33% of the CDC members were women. A vast majority (98%) of the elected CDCs received block grants for project implementation. Repeat block grants were introduced in NSP III and reached about 157 (93%) districts. Close to 89,000 CDC projects were financed. The total disbursement in block grants was more than US$1.6 billion. Nearly 81,000 CDC projects, financed by NSP, were completed and benefitted 13 million Afghans.

Linkages of results with global development goals

Funded projects were generally in the following categories:

- Transportation and communication (village pathways, bridges and secondary roads facilitating mobility of villagers, connecting once isolated villages to each other and providing access to markets, health, and education facilities);
- Water and sanitation (latrines, public baths, handpumps for drinking water, water-divider construction, pipe schemes, and aqueducts);
- Utilities (electricity supply) in the remotest rural areas;
- Livelihood and income generation (irrigation facilities for small farmers to improve food production, income and employment generation in CDC-led projects, and provision of literacy and vocational training);
- Other constructions include construction and renovation of health facilities and schools.

The central themes of these projects address, directly and indirectly, MDGs and SDGs. Certain emerging features generated through CDC project actions provide grounds for celebrating the success of NSP in serving the poor, showcasing how local actions can spur development and help achieve the global development goals at local levels. The World Bank's and UN agencies' reports bear evidence that NSP made a substantial and sustainable contribution to the country's infrastructure and socio-economic development through positive changes brought about at local levels. Communities in villages, where CDCs are operational, perceive CDC project results as tangible and beneficial for their communities.

CDC projects providing improved access to infrastructure (irrigation, power, transportation, water supply, and other services) helped promote rural communities' overall well-being, indicating successful local actions on SDG 3 – wellbeing for all at all ages.

CDCs' local actions promoted better health conditions by improving access of communities to clean drinking water and sanitation facilities, both integral to healthy living. These results directly link to SDG 6 (clean water and sanitation) contributing to good health (SDG 3). In total, over 9,700,000 rural poor have had access to drinking water and safe sanitation facilities through NSP financing.

CDC projects promoted healthcare indirectly also, through support given in other service sectors, such as transportation and communication, specifically, road construction. Roads promoted accessibility to health facilities and helped health workers bring counselling and minimum healthcare at the doorsteps of rural households, especially needed by vulnerable child patients and prenatal and postnatal women encountering mobility issues. To respond to the heath needs of communities, a limited number of health facilities (89 health clinics and 9 hospitals) were constructed or rehabilitated by CDCs during the NSP implementation period.

CDCs made contributions to the education sector – MDG 2 (universal primary education) and SDG 4 (quality education). The roads, by connecting villages to schools, are known to have increased girls' school attendance. Construction and renovations of education facilities have also been funded by NSP. Over 6,400 classrooms and 5,800 community centres, for learning cum socialization, have been constructed or renovated.

Access to modern energy services, such as electricity, is considered essential for improving people's quality of life and thereby people's well-being. There is evidence that use of electricity has been rising in the NSP villages. Basic electricity generated, through solar, micro-hydro, biogas or wind, provides affordable clean energy, addressing SDG 7, while also representing, albeit in a very limited way, local level climate action (SDG 13) for generating conditions for a less polluted environment and thus better health conditions (SDG 3). Over 4.2 megawatts of electricity have been generated through micro-hydropower plant construction and gridline extension.

Infrastructure projects, such as irrigation facilities for small farmers, represent actions at the lowest local unit (villages) for improving food production to reduce hunger (MDG 1 – eradicate hunger and SDG 2 – zero hunger). With NSP financing, irrigation brought more areas under cultivation and helped increase agricultural yield. CDC-financed irrigation facilities irrigated 1,300,000 acres of agricultural land with construction and rehabilitation of 4,400 downstream irrigation canals.

In conformity with the gender equality component of the global development goals (MDG 3 – gender equality and empowerment of women and SDG 5 – gender equality), NSP embraced the concept of the equal role of men and women in the society. CDCs are mandated to give equal opportunities to men and women by ensuring that women have equal shares of leadership positions in the CDCs. Half of all CDC seats are allocated to women to ensure women members' equal share of benefits from the funds received by CDCs. According to sample surveys, close to 50% of the NSP beneficiaries are women. Of the women surveyed, close to 80% take active roles in the CDCs. The NSP operation has definitely made a substantial contribution to an increased and more meaningful role of women in decision-making processes, conforming to the global agenda of gender equality. Female CDC members surveyed say that broader improvements have been brought to women's lives through participation in CDC activities (Echavez, 2012).

NSP also provided employment and income opportunities in alignment with SDG 8, addressing the need for decent work. The block grants of NSP generated over US$66 million paid labour days for contracting local skilled and unskilled workers for implementing the CDC projects, thus generating income and employment for the rural poor. NSP-introduced Maintenance Cash Grant Program ensured that the employment generated would provide an estimated two to three months of food security to participating households.

Under NSP, progress was achieved in using CDCs for coordinating service delivery in local areas, on behalf of ministries, other than the MRRD that housed NSP. Ministry of Agriculture worked with CDCs to deliver agriculture extension services. CDCs in remote areas helped a number of Provincial Departments of Education in school construction.

The benefits out of CDC projects, funded by NSP, made contributions to the generation of national-level results addressing the global development goals, even under deteriorating security and economic conditions in the country. Healthcare coverage at the national level expanded, and national maternal and infant mortality rates declined. The population's access to water, electricity, and road connectivity significantly improved. According to the Afghanistan government's calculations, in the period 2007 to 2014 access to safe drinking water increased from 27 to 65% of the national population. In the same period, access to electricity increased from approximately 42% to 90% of the nation's population, indicating a major gain. Undeniably, CDC projects, covering all provinces in the country, made contributions to these advances at the national level, amidst deteriorating security and economic decline.

Better spending

These gains have been achieved at lower costs than donor financed projects, implemented by donor contracted executing agencies, engaging expensive expatriate consultants, and charging high overhead costs of up to 15% or more of the total budget of projects, and a higher percentage as security costs. On the other hand, NSP has had tangible deliverables at a price tag considerably lower than large-scale Western government– or international NGO–led initiatives. Spending to effectively address people's priority needs at a low cost represents better spending.

Many examples of projects with huge price tags but flimsy results are found. A US$280 million project, Promote, financed by the United States for women's development in Afghanistan, was a waste of money. The objective was to help 75,000 women to get jobs and promotions through apprenticeships and internships (Nordland, 2018). It is reported that only 55 women have been promoted to better jobs; and it is not clear if the Promote project could be credited for these promotions. The complaints of Afghan women activists are that much of the money had been spent on administrative costs and hiring American contractors. A report of the American Special Inspector General for Afghanistan Reconstruction (SIGAR) states that his inspection after five years of the start of this project showed that three expatriates' security and overhead costs accounted for 18% of the US$89 million disbursed. Yet another project, Dahla Dam, funded by Canada for US$50 million, was a failure due to design and other flaws (Watson, 2012). Afghan residents in the area say that too much money was lost to indirect costs, especially payments to a private security firm. Afghan officials familiar with the project say that most funds were spent on contractual payments to SNC Lavalin, contracted by the Canadian government, their housing, home leave travels, and security. SNC Lavalin retained highly paid Canadian staff, occupying two palatial villas in Kabul and Kandahar. The infamous private security firm Watan Risk Management Group was contracted. The firm allegedly paid protection money to the Taliban. Security costs reached US$10 million, 20% of the total budget.

On the other hand, CDCs, using local labour and material, can get projects completed at a much lower cost. NSP's delivery cost was lower also because of considerable investment of time and resources by community members into the projects, of which the communities take ownership. The NSP regulations required the community members to contribute at least 10% (in cash or in kind) of the total cost of each project. In addition to addressing this mandatory requirement, the villagers are known to input hundreds of volunteer hours into every stage of a project from identification and planning to implementation, monitoring and reporting, contributing to low-cost project delivery. Besides, NSP spent little on securing the project sites. The word is that in acknowledgement of the popularity of the NSP in rural communities of Afghanistan, the Taliban avoided attacking its projects. Also, villagers' engagement at every stage of a project, from identification and planning to implementation and monitoring, nurtures a sense of ownership of these projects in the communities. This spirit of ownership leads them to protect the projects with life against any onslaught.

All of these factors make CDC-delivered projects cost beneficial. According to World Bank calculations, NSP had a very high economic rate of return, which well meets better spending criteria. The World Bank measured the cost efficiency of NSP-financed projects, including overheads and facilitating partners' CDC facilitation costs. The economic rate of return is found to be highest, at 122% in water supply projects, followed by 70% in roads and bridges; 42% in energy; and 17% in irrigation. The average rate of return combining all these sectors is impressive – 63%, a very significant indicator of better spending for development.

Start-up of NSP succession era

After running for 13 years (2003–2016) NSP closed in March 2017 and has been succeeded by Citizens' Charter Afghanistan Project (CCAP), which will build on the successes of NSP and also address some gaps (World Bank, 2018).

CCAP is one of the components of Citizens' Charter, a National Priority Program of the government of Afghanistan. The Citizens' Charter represents a social contract between the Afghan state and the citizens for delivery of the needed services. CCAP's mandate is to help the Afghan government in fulfilling its commitment to the citizens for delivery of basic services to all, based on community prioritization of the services. CCAP plans to reach one-third of the country in all 34 provinces, an estimated ten million people (Loha, 2018). In the first four years, approximately 8,500,000 people will be reached. The aim, as under the NSP in the past, is to reduce poverty and raise citizens' living standards by improving the delivery of core infrastructure services in rural areas – roads, access to schools, health facilities, drinking water, small irrigation, and electricity – through the CDCs. The participatory community-driven development approach continues.

The CDCs, the local institutions built by NSP in the first decade-and-a-half of the new millennium, are continuing their operation under the CCAP mandate: to nurture the CDCs and strengthen and expand them to improve the delivery of infrastructure and other services to the people. CDCs serve as building blocks for nationalization of the SDGs. The CDCs, continuing service delivery to meet the socio-economic needs of the poor through community-driven development actions, enable the Afghan government to nationalize many of the global development goals included in the 2030 Agenda.

An expanded mandate

To strengthen the Afghan government's arm for service delivery, certain expansionary measures have been taken. CCAP is to operate in both rural and urban areas (Government of Islamic Republic of Afghanistan, 2016). Along with the rural community councils serving the rural poor, urban Community Development Councils are formed for crucial service delivery to the urban poor. Urban block grants from CCAP are used to implement projects for access of the urban poor to potable water (SDG 6, clean water and sanitation) as well as electricity and lighting (SDG 7, affordable clean energy), and to promote women's livelihoods, addressing SDG 5 – gender equality. Notably, under NSP, CDC projects established a good track record of enhancing the decision-making power of women and thus promoting gender equality at a low cost when compared to large and expensive external donor-financed projects. A project, promoting Women's Political Participation in Afghanistan, financed by Canada at US$5,600,000, had an extremely limited impact in changing women's capacity to influence decision-making processes, according to an evaluation report.

Given that in Afghanistan, cities are growing at a rapid rate and the urban poor – an estimated 29% of the urban population – are living in slum-like conditions, with little access to basic services, CCAP focuses on SDG 11, sustainable cities and communities. The urban area projects concentrate on upgrading roads and sidewalks and improving drainage, as well as introducing solid waste management and building parks, playgrounds, and green spaces, all linked to the theme of SDG 11, sustainable cities. Services in urban areas will also focus on healthcare (SDG 3, good health and well-being) and education (SDGs 4, quality education).

A measure that will make service delivery, across the country, more efficient, better targeted, and cheaper is consolidation of vital service delivery from several ministries – namely, agriculture and livestock (MAIL), Rural Rehabilitation and Development (MRRD), health (MoPH), and education (MOE), under the aegis of CCAP. Service delivery functions of these ministries will be coordinated by CCAP through the CDCs. For instance, agriculture services of MAIL in provision of input packages for production of wheat and horticulture products, rehabilitation of irrigation networks and on-farm water management, conservation of natural resources, and provision of veterinary services for livestock will be managed by CCAP CDCs. These services generate income and contribute to economic development at the community level in the immediate term and at the national level in the long term.

Transferring resources and responsibilities for local level management of central ministries' services bear enormous potential. Delivery through the CDCs makes service delivery and monitoring simple and efficient, better targeted and cost beneficial. It is argued that community-managed projects are more accountable and less corrupt, as CDCs tend to nurture a sense of ownership of the projects they manage. CDC members' extensive and in-depth knowledge of local issues and the terrain in which the projects operate promote easier and efficient procurement and project implementation. CCAP, assuming the coordination role, is able to ensure appropriate targeting of projects for the poor and the vulnerable and for promoting women's role as decision-makers.

As argued earlier, delivery costs of community-led projects are lower than those implemented by donors because of the use of locally available material, supplies, and labour and the volunteer hours contributed by community members who take ownership of projects. The same logic applies to community-led versus nationally led projects. The coordination role assumed by CCAP for delivering services for the central ministries is considered to be a major cost saving factor reflecting better spending. The average internal rate of return for service delivery by CDC projects, on behalf of the central ministries, is expected to be high, well above global averages.

NSP had overreached the planned targets for both outputs and outcomes. The World Bank issued progress report of 2019 shows that CCAP implementation is going well in terms of outputs. Progress towards the outcomes can only be rated as fair, given the security conditions in the country. In rural areas, over 8,400 new CDCs have been elected; close to 7,000 Community Development Plans (CDPs) have been drawn up; and over 5,700 project proposals have been completed. In the urban areas, CCAP has been extended to 666 communities; over 600 CDCs have been elected; close to 600 CDPs (Community Development Plans) have been completed; and over 500 projects have been approved for funding. Over the years, the CDCs have and will continue to address the priority policy objectives of the Afghan government to align the country's development to the global development goal of multidimensional poverty eradication, providing low-cost solutions to seemingly intractable problems.

Following better spending formulae, Afghanistan's community-driven development programme marches forward

Working across the nation, in all 34 provinces, the CDCs are continuing to deliver and improve rural infrastructure – construct and rehabilitate roads, bridges, irrigation canals, schools, and health facilities; install hand pumps and sanitation facilities and the necessary infrastructure for electricity supply – opening up opportunities for hundreds of thousands of the rural poor, and now, the urban poor under the CCAP initiative. Opportunities for basic education, healthy living, and improved access to food are promoted. More than US$80 of labour for the CDC infrastructure projects comes from the communities themselves, keeping costs low, while generating wages (employment) for the poor. Community involvement carries the added advantage of generating a sense of ownership, which provides incentives to provide appropriate maintenance of the infrastructure created. The highest positive sustained impact has been on gender equality. The female CDC members are happy with their participation in the CDCs as they feel that they have been given equal decision-making power in their community's development councils.

The benefits came with a low price tag, indicating cost efficient investments, even when the CDCs have been operating in a deeply fragile and conflict-affected situations in the country. Albeit limited by the problematic context, the community-driven development approach of the NSP and the follow-up Citizens' Charter Afghanistan Program (CCAP) support the Afghan government's efforts to nationalize the global development goals. NSP is the Afghan government's flagship programme for rural poverty reduction. Along with pursuing the cause of rural poverty reduction, the CCAP is also addressing urban poverty. Both programmes are based on recognition of the autonomy of local communities. Investments in such programmes help achieve a better future for all by addressing the global challenges posed by poverty and inequality. Despite security threats posing formidable challenges, the CDCs of NSP and CCAP continue to pursue the agenda of key development services delivery to the poor that is also central to the 2030 Agenda of *leaving no one behind.*

References

Beath, A., Christia, F., & Enikolopov, R. (2015) The national solidarity programme: Assessing the effects of community-driven development in Afghanistan, *International Peacekeeping*, 22(4), pp. 302–320.

Centre for Public Impact. (2016) Building trust in government: Afghanistan's National Solidarity Programme (NSP). Available at: www.centreforpublicimpact.org/case-study/building-trust-in-government-afghanistans-national-solidarity-program/

Echavez, C. R. (2012) *Does Women's Participation in the National Solidarity Program Make a Difference in Their Lives*. Kabul, Afghanistan Research and Evaluation Unit (AREU).

Government of Islamic Republic of Afghanistan. (2016) *Citizens' Charter National Priority Programme*. Kabul, Government of Islamic Republic of Afghanistan.

Guggenheim, S., Wiranto, T., Prasta, Y., & Wong, S. (2004) Indonesia's Kacamatan development program: A large-scale use of community development to reduce poverty, in *Scaling up Poverty Reduction: A Global Learning Process and Conference*. Shanghai, World Bank, pp. 1–26.

Loha, W. (2018) Citizens' charter Afghanistan project (CCAP). Available at: https://community leddev.org/2018/02/12/citizens-charter-afghanistan-project-ccap/ [Accessed 11 April 2019].

Nordland, R. (2018) U.S. aid program vowed to help 75,000 Afghan women: Watchdog says it's a flop, *New York Times*, 13 September. Available at: www.nytimes.com/2018/09/13/ world/asia/afghanistan-women-usaid.html

Sadri, A. (2019) Achieving accountability through Afghanistan's community development program, 11 May. Available at: www.democracyspeaks.org/blog/achieving-accountability-through-afghanistan%E2%80%99s-community-development-program [Accessed 10 April 2019].

Watson, P. (2012) Canada's Afghan legacy: Failure at Dahla Dam. *Toronto Star*, 14 July. Available at: www.thestar.com/news/world/2012/07/14/canadas_afghan_legacy_failure_ at_dahla_dam.html

World Bank. (2018) Flagship Afghan rural program lays strong foundation for the future. Available at: www.worldbank.org/en/news/feature/2018/01/16/flagship-afghan-rural-program-lays-strong-foundation-for-future

World Bank. (2019) Supporting inclusive growth in Afghanistan. Available at: http://projects-beta.worldbank.org/en/results/2019/04/01/supporting-inclusive-growth-in-afghanistan

Further reading

Barron, P. (2011) *Community Driven Development in Post Conflict and Conflict Affected Areas: Experiences from East Asia*. Washington, DC, World Bank.

Bhatia, J., Jareer, N., & McIntosh, R. (2018) Community-driven development in Afghanistan: A case study of the national solidarity programme in Wardak, *Asian Survey*, 58(6), pp. 1042–1065.

Boesen, I. W. (2004) From subjects to citizens: Local participation in the national solidarity program. *The Afghanistan Research and Evaluation Unit (AREU) Working Paper Series*. Kabul, AREU.

Brick, J. (2008) Investigating the sustainability of community development councils in Afghanisthan. *Unpublished Report Prepared for JICA*. Kabul, AREU.

Calder, J., & Hakimi, A. (2009) *Statebuilding and Community Engagement Without Reconcilation: A Case Study of Afghanisthan's National Solidarity Program*. New York, Future Generations Graduate School.

Couch, G. (2013) World Bank releases randomized impact evaluation of Afghanistan's national solidarity programme. Available at: www.worldbank.org/en/news/feature/2013/ 09/29/world-bank-releases-randomized-impact-evaluation-of-afghanistans-national-solidarity-programme

Government of Afghanisthan. (2017) *Voluntary National Review at the High Level Political Forum SDGs' Progress Report Afghanistan*. Kabul, Government of Afghanisthan.

Humayun, A. A., Exum, A. M., & Nagl, J. A. (2009) *A Pathway to Success in Afghanistan: The National Solidarity Program*. Washington, DC, Centre for New American Security.

Khedir, T. (2017) Electricity connects Afghan village to wider world. Available at: www. worldbank.org/en/news/feature/2017/02/13/electricity-connects-afghan-village-wider-world

Majeed, R. (2014) *Building Trust in Government: Afghanisthan's National Solidarity Program 2002–2013*. Princeton, NJ, Princeton University.

Participedia. (2018) Afghanistan National Solidarity Program. Available at: https://participedia. net/en/cases/afghanistan-national-solidarity-program

UN Habitat. (2017) *Analytic Closure Report: National Solidarity Programme (NSP)*. Kabul, Afghanistan, UN Habitat.

Wong, S., & Guggenheim, S. (2005) Community-driven development: Decentralization's accountability challenge, in *East Asia Decentralizes: Making Local Government Work*. Washington, DC, World Bank, pp. 253–267.

World Bank. (2013) National solidarity program III. Available at: http://projects.worldbank. org/P117103/national-solidarity-program-iii?lang=en

World Bank. (2017) *Atlas of Sustainable Development Goals: From World Development Indicators*. Washington, DC, World Bank Group.

Yemak, R., Gan, F., & Cheng., D. (2013) *Celebrating Ten Years of the National Solidarity Program (NSP): A Glimpse of the Rural Development Story in Afghanistan*. Washington, DC, World Bank.

10 Multidimensional poverty measurement and aid efficiency

A case study from Afghanistan

Abdullah Al Mamun and Sanni Yaya

The article will present a poverty measurement tool named the Programme Area Multidimensional Poverty Index (PAMPI) by the senior managers of the development agency to measure the long-term outcome of programmes and to increase aid efficiency in Afghanistan. Increasingly, academics and development practitioners are taking a multidimensional view of poverty while proposing interconnected, cross-cutting solutions. The movement towards using the multidimensional approach as a policy tool for poverty reduction has accelerated rapidly; most notably the United Nations Development Program (UNDP) has developed the Multidimensional Poverty Index (MPI) for developed and developing countries and has included it in its Human Development Report since 2010. At the country level, the MPI methodology has been adopted by several countries to develop better anti-poverty policies and to increase the efficiency of financial expenditure. Similarly, this approach could be applicable at the institution level, and international and national development agencies could use this MPI methodology as an effective and efficient tool to evaluate the impact of aid in poverty-reduction programmes in meeting the Sustainable Development Goals (SDGs).

Keywords

Poverty measurement, aid effectiveness, poverty in Afghanistan, Multidimensional Poverty Index, sustainable development

Background

Increasingly, academics and development practitioners are taking a multidimensional view of quality of life (QoL) and poverty while proposing interconnected, cross-cutting solutions (OPHI, 2019a). Writing on QoL, the Stiglitz-Sen-Fitoussi Commission on the Measurement of Economic Performance and Social Progress put it bluntly: "(W)hen designing policies in specific fields, impacts on indicators pertaining to different quality-of-life dimensions should be considered jointly, to address the interactions between dimensions and the needs of people who are disadvantaged in several domains" (Stiglitz et al., 2009, 15). The movement towards a multidimensional approach has accelerated rapidly; most notably the UNDP has developed the Multidimensional Poverty Index (MPI) for 105 countries and has included it in its Human Development Report since 2010 (UNDP, 2018). At the country level, the MPI methodology adopted here has been used by Bhutan, Brazil, China, Colombia, and Mexico, to name only a few examples (MPPN, 2019). In the case of Columbia, the government announced a plan to reduce the rate of poverty of the entire population by 22% (from 30% to 8%) by 2030 based on their country -pecific Multidimensional Poverty Index (Colombia, 2019). Most recently, the government of Afghanistan launched its first MPI on March 31, 2019. The report revealed that approximately 52% of the population is multidimensionally poor. Considering provincial poverty status, the rate of poverty is the lowest in Kabul, 14.7%, and the highest in Bagdish, 85.5% (National Statistics and Information Authority, 2019)

Considering the potential utility of an indexed measure for decision-makers at SMEC Foundation Afghanistan (SFA), the senior managers decided to develop a Programme Area Multidimensional Poverty Index (PAMPI) to measure and increase aid efficiency in reducing poverty. The index is a robust evaluation tool to measure the poverty status of people within programme areas. It is an index that identifies multiple deprivations at the level of an individual or household across several dimensions (individually composed of multiple indicators). Here, a household is considered to be poor depending on the number of deprivations that they experience across multiple dimensions. The class of Multidimensional Poverty Index proposed here uses a dual cut-off criteria to identify the poor on each dimension, as well as across dimensions.

Purpose

The objective of the PAMPI is to deliver a new poverty measurement tool named as Multidimensional Poverty Index to senior managers to measure the long-term outcome of programmes and to increase aid efficiency in Afghanistan. In the past, programme managers utilized a "dashboard" approach to report the status of a wide number of indicators separately. While this approach has the virtue of allowing decision-makers to "drill down" to a single indicator for closer inspection, it also suffers from a number of serious limitations:

- It provides no insight into which indicators are important. For example, a typical survey might collect data on 25 separate indicators having to do with poverty; identification of a single indicator or group of indicators to assess the level of poverty is entirely a subjective decision left to the individual. In other words, to answer the question "What is the level of poverty?" a manager must decide which of the 85 indicators they think are most important. This is not a problem when a single, well-defined indicator is sufficient as the basis for measurement (e.g., income), but this practice is increasingly regarded as overly reductionist and unable to accurately reflect complex realities.
- The interconnected deprivations that makeup people's experience of poverty are often overlooked by a one-dimensional measurement of poverty such as

income poverty. This is particularly obvious in a country like Afghanistan where data on cash income is both scarce and, in any case, captures only a fraction of the normative understanding of what it means to be poor.

- A dashboard approach does not help to understand the joint distribution of deprivations that constitute poverty even when managers take multiple indicators into consideration. For example, imagine two communities (A and B) with the same deprivation rate in education (70%), health (60%), and food security (40%) indicators. Using a dashboard approach, a narrative report would describe that "70% of households do not have access to education services; 60% of households do not have access to health services; and 40% of households do not have food security in both communities A and B." Here, the report would correctly note the deprivation in each separate indicator but would say nothing about the joint distribution of the indicators (e.g., Does the same household suffer simultaneously from one deprivation, two, or all three?).

This latter point is illustrated in Figure 10.1, which illustrates the importance of measuring the joint distribution of deprivations. In community A, many of the households that are deprived in one indicator are also deprived in at least one other indicator. In community B, the situation is considerably different: only a few households are deprived in multiple indicators (most suffer from a single deprivation). If a higher rate of poverty means that households suffer from multiple deprivations, then the proportion of households who suffer from poverty is considerably greater in community A than in community B, even though they have the same deprivation rates in all indicators separately.

The senior managers at SFA are interested in gaining a holistic insight into the poverty status of beneficiaries as well, one that does not only include the dashboard approach. As a result, working with academics and practitioners, SMEC Foundation developed its own PAMPI.

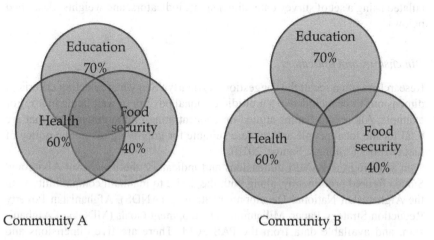

Community A   Community B

Figure 10.1 Comparison in the joint distribution of deprivations in communities A and B

The PAMPI value is a poverty measure with a high-resolution lens (Alkire, 2010). In other words, it provides a single, overall measure of the proportion of people who are multidimensionnaly deprived. Besides, this measure shows the intensity of poverty. At the same time, it allows senior managers to focus on discreet components of the PAMPI value, broken down according to population subgroups (e.g., by gender, ethnicity, age group, and other sociodemographic characteristics of interest), region, dimension, and indicator. This characteristic of PAMPI provides SFA senior managers with both a quick, executive summary of overall poverty and the policy-specific data required for a more fine-grained analysis of individual poverty dimensions and indicators.

Measuring PAMPI

The PAMPI is based on the data from the Program Area Survey (PAS) 2014, conducted by the SFA in six provinces (Bamyan, Parwan, Baghlan, Samangan, Takhar, and Badakhshan) where SFA has a significant programme presence. The first survey was conducted in 2014 and will be repeated every five years in order to provide comparable time-series data (the next PAS will be conducted in 2019). In each region, districts were selected to be included in the PAS if the security situation was regarded as stable and where two or more SFA sectors (excluding infrastructure) were active. There where 30 districts selected for the PAS 2014, home to approximately 1.6 million people (CSO, 2014). A list of the districts selected in each region is provided in Appendix A.

PAMPI methodology

The PAMPI uses the mathematical structure of the Alkire-Foster poverty measurement methodology (or AF methodology). The values of the PAMPI are calculated using a set of survey data, dimensions, indicators, and weights (described below).

Dimensions and indicators

Researchers often recall the suggestion of Amartya Sen when selecting criteria of dimensions to be included in a multidimensional poverty or well-being index. For example, Alkire and Santos argued on concentrating on the dimensions that are vital for the local people and that are suitable for improving the public policy of that territory (Alkire & Santos, 2010).

In selecting the PAMPI dimensions and indicators, the insights of Alkire and Santos figured prominently, along with the desire to maintain compatibility with the Afghanistan National Development Strategy (ANDS), Afghanistan Poverty Reduction Strategy Paper, Millennium Development Goals (MDGs) of Afghanistan, and available data from the PAS 2014. There are five dimensions and 14 indicators included in the PAMPI. The indicators are predominantly related to MDG 1: eradicate extreme poverty; MDG 2: achieve universal primary education; MDG 5: improve maternal health; MDG 7: ensure environmental sustainability; and MDG 8: develop global partnership for development.

The PAMPI values are calculated using a data set of households. In the calculation process, every household is defined as "deprived" or "not deprived" in each indicator using a researcher-defined deprivation cut-off. The deprivation cut-off for each indicator was chosen based on a review of the available literature and discussion with experienced development practitioners who are familiar with the Afghanistan context. Table 10.1 gives a breakdown of the deprivation cut-off points for each indicator, and also shows their relationship with the MDGs.

Weights of dimensions and indicators

In multidimensional poverty measures, weights can be applied among dimensions (e.g., distribution of weights among food security, employment, education, and so on), among indicators of a dimension (e.g., allocation of weights among the indicators – electricity, water, sanitation, and so on – of living standard dimension), and among homogeneous groups of people (Alkire & Santos, 2010). In the case of the PAMPI, all five dimensions are of considerable importance for a post-war country such as Afghanistan; however, it is very difficult to measure the relative importance of one dimension in comparison to other dimensions. In the PAMPI, all five dimensions are weighted equally and all the indicators within a dimension (in case of more than one indicator) are also weighted equally. This process is in line with existing practice in other multidimensional poverty indices (OPHI, 2019b).

To check for robustness and to assess the variation of the PAMPI value, the study considered three measures, in consultation with the SFA managers: Measure 1 is composed of five dimensions and 15 indicators with equal weight for every indicator in each dimension; Measure 2 is composed of five dimensions and 14 indicators with equal weight for every indicator in each dimension; Measure 3 is composed of three dimensions and ten indicators with equal weight for every indicator in each dimension – an approach that utilizes a similar number and type of dimensions and indicators used by the UNDP to develop their MPI for more than 100 countries. Table 10.2 shows the dimensions, indicators, and weights used for each measure.

After careful consideration, Measure 2 was deemed most appropriate as a standard poverty measure for the SFA because it contains all five dimensions and all 14 indicators with less than 1% missing values. Measure 1 also contains five dimensions, but there is one indicator, "nutrition," that represented a total of 16.1% missing values. A list of indicators with missing values is provided in Appendix B.

While this researcher-led approach is sufficient for the needs of SFA, we do recognize the benefits of adopting a more participatory approach, one that enables researchers to develop more locally specific poverty measures based on a shared social understanding of what it means to be in poverty (Mitra et al., 2011; Ravallion, 2011). Although this more participatory approach is not presented here, SFA will run focus group discussions in the future, depending on available resources, to further develop a culturally grounded poverty measure that addresses this limitation.

Table 10.1 Dimensions, indicators, deprivation cut-offs, and relationships with MDGs

Dimension	Indicator	A household is deprived if…	Relationship with MDG
Food security	Food security	During the last 12 months, the household struggled (self-reported) to have enough food	MDG 1
Employment	Employment	Household does not have at least one employed adult (age 19–65) member out of every six household members. If the household has less than six members and no one is employed, then the household is also deprived.	MDG 1
Health	Maternal and child health	The household has delivered a baby in the last five years without the help of a skilled birth attendant (nurse/midwife/doctor) or trained traditional birth attendant. If the household has not delivered a baby in the last five years, then the household is not deprived.	MDG 5
Education	Years of schooling	No household adult (age 19 and over) member in the household has completed at least a primary education	MDG 2
	Children enrolment	Any school-aged (7–15) child from the household is not enrolled in school. The household is not deprived if there is no school-aged child living in the house.	MDG 2
Living standard	Electricity	The household has no electricity	MDG 7
	Sanitation	The household does not have a private pit latrine or shared/public pit latrine as sanitation facility	MDG 7
	Water	The household does not have access to clean drinking water, or clean water is more than 30-minute walk from home in the cold season or warm season	MDG 7
	Cooking fuel	The household does not have sufficient firewood, kerosene/oil, electricity, gas, or solar as a source of cooking fuel	MDG 7
	Heating source	The household does not have sufficient gas, wood, kerosene, or electricity as the main source of heating for the winter	MDG 7
	Asset	The household does not have at least one of the following assets: radio, tape recorder, TV, DVD/VCD/CD/VCR, telephone, bike, sewing machine, or electric fan	MDG 7
	Floor cover	The household does not have more than one type of floor-covering carpet	MDG 7
	House ownership	Any household member does not own a house for living	MDG 7

Table 10.2 Dimensions, indicators, and weights used for the measures

Dimension	Indicator	Measure 1	Measure 2	Measure 3
Food security	Food security	1/5	1/5	
Employment	Employment	1/5	1/5	
Education	Year of schooling	1/10	1/10	1/6
	Children enrolment	1/10	1/10	1/6
Health	Nutrition	1/15		
	Maternal and child health	1/15	1/10	1/6
	Prenatal care	1/15	1/10	1/6
Living standards	Electricity	1/40	1/40	1/18
	Sanitation	1/40	1/40	1/18
	Water	1/40	1/40	1/18
	Cooking fuel	1/40	1/40	1/18
	Heating source	1/40	1/40	1/18
	Floor cover	1/40	1/40	1/18
	Asset	1/40	1/40	
	House ownership	1/40	1/40	

Multidimensional poverty cut-off

The value of the poverty cut-off is used to define whether a household is multi-dimensionally poor or not. Defining a poverty cut-off is a not an objective process, and may vary from one country to another, between socio-economic groups, between age groups, and so on. For the purpose of this study, in consultation with the senior managers, we considered 40% as the poverty cut-off; this means that a household in any SFA programme area is considered to be *multidimensionally poor* if the household is 40% deprived of weighted indicators. The levels of poverty for different cut-offs are provided in Appendix C.

As an example, in the PAMPI there are five dimensions with equal weight of 0.2 or 20% for each. The dimension *food security* contains only one indicator, which is also *food security*. Therefore, the weight for the indicator *food security* will be 20%. Similarly, the indicator *employment* is also weighted at 20%. Each of the indicators for the remaining dimensions is weighted differently: each indicator of the *education* dimension holds 0.10 or 10%; indicators in the *health* dimension also hold 10%; and indicators in the *living standard* dimension hold 0.025 or 2.5% of the overall 100% weight. Accordingly, a household is considered to be poor if it is deprived in at least the following:

- all the indicators of any two dimensions out of the five dimensions;
- the food security or employment indicator, any education or health indicator, and any four living standard indicators;
- any possible combination of indicators that result in a cumulative deprivation equal to or greater than 40% out of the 100% weight.

Results

The PAMPI is a product value of the incidence of poverty (H) and the average intensity of poverty (A) in the coverage area at the household level. The incidence of poverty is defined as the proportion of the households that are multidimensionally poor, and the intensity of poverty is defined as the proportion of indicators that fall below the deprevation cut-off.

Table 10.3 shows the name of the survey in the first column and the year of conducting the survey used to measure the PAMPI in the second column. It reports in the next five columns the values of the overall PAMPI, the incidence of poverty (H), the average intensity across the poor (A), the percentage of the population vulnerable to poverty, and the percentage of the population in sever poverty. The PAMPI value (0.42) is calculated by multiplying the values of the incidence of poverty (H) and the average intensity across the poor (A) – that is, PAMPI = HxA. The table shows that 71.6% households are deprived in at least 40% of weighted indicators. The sixth column of the table reports the proportion of households identified as "Vulnerable to poverty," defined as households deprived in 30% – –40% of the weighted indicators. The last column reports the percentage of households identified as in "Severe poverty," defined as being deprived in 50% or more of the weighted indicators.

Incidence of deprivation in each of the PAMPI indicator

There are 14 indicators considered for computing the PAMPI in five dimensions: food security, employment, education, health, and living standard. The following chart (Figure 10.2) reports the deprivation of poor people in each indicator. The deprivation of non-poor people is not included here.

Figure 10.2 shows that most of the poor households are deprived (more than 60%) in food security, cooking fuel, and heating source. A significant proportion, approximately 50%, of poor households are also deprived in years of schooling, maternal and child health, electricity, sanitation, and drinking water indicators. In comparison to other indicators the level of poor households' deprivations are less in children enrolment and asset and house ownership indicators.

Table 10.3 Programme Area Multidimensional Poverty Index (PAMPI)

Survey	Year	Programme Area Multidimensional Poverty Index (PAMPI = H×A)	Incidence of poverty (H)	Average intensity across the poor (A)	Percentage of population vulnerable to poverty	Percentage of population in severe poverty
Programme area assessment	2014	0.42	71.6%	58.4%	14.7%	52.6%

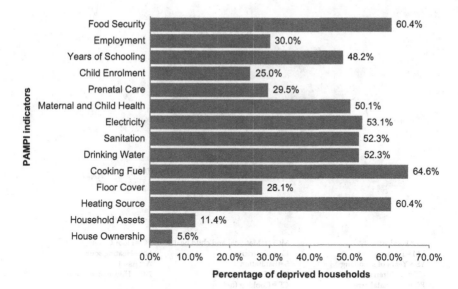

Figure 10.2 Deprivations in each indicator

Contribution of indicators to PAMPI value

The PAMPI is a calculated value of weighted indicators. The contribution of each weighted indicator to the PAMPI value can be broken down by percentage.

The pie chart (Figure 10.3) below shows the proportion of the weighted indicators that contribute to the PAMPI. The contribution of food security (FS) is the highest and employment (EM) is the second highest in terms of the indicators' contributions to the PAMPI value. The contributions of maternal and child health (MCH) and years of schooling (YS) are also significant: 12% and 11.7% respectively. The analysis suggests that SFA should prioritize efforts to improve food security, employment generation, and maternal and child health within programme areas in order to reduce the poverty level of its beneficiaries.

Intensity of PAMPI

As previously reported, a household is considered poor if it is deprived of 40% of weighted indicators. The following figure (Figure 10.4) illustrates the proportion of households that are poor on the basis of their relative weighted indicators.

Figure 10.4 can be read as follows: the first column, labeled as "40%," indicates that 71.6% of Afghan households are deprived in 40% or more of the weighted indicators. The second bar shows that more than a half of the households (52.6%) in SFA programme areas are deprived in 50% or more of weighted indicators. Similarly, the third bar shows that approximately one-third of households in SFA programme areas are deprived in 60% or more of the weighted indicators. The last bar shows that there is no household deprived in all the indicators of PAMPI.

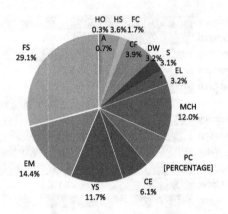

HO   HS   FC
0.3% 3.6% 1.7%
A
0.7%   CF   DW   S
3.9% 3.2% 3.1%

FS
29.1%

EL
3.2%

MCH
12.0%

PC
[PERCENTAGE]

EM
14.4%

CE
6.1%

YS
11.7%

FS = Food security   MCH = Maternal and child health   FC = Floor Cover
EM = Employment   EL = Electricity   HS = Heating sources
YS = Years of schooling   S = Sanitation   A = Asset
CE = Children enrolment   DW = Drinking water   HO = House ownership
PC = Antenatal care   CF = Cooking fuel

Figure 10.3 Contribution of indicators to the PAMPI

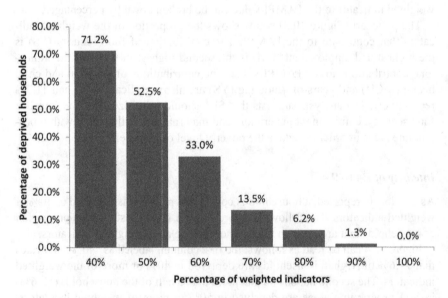

Figure 10.4 Households' deprivation in 'X'% or more of the weighted indicators

PAMPI at the regional level

The PAMPI can be broken donw according to SFA regions in order to highlight the differences between them. Table 10.4 below shows the PAMPI values, disaggregated according to SFA programme regions.

Table 10.4 shows in the second column that Badakhshan has the highest PAMPI value, suggesting that the poverty situation in Badakhshan is the worst among all the four SFA programme regions. Bamyan and Takhar are slightly better off, but they are still very similar to Badakhshan's PAMPI value. The third column shows that 75.8% of households are multidimensionally poor in Badakhshan, which reflects the highest incidence of poverty among the SFA programme regions. The fourth column reports that the intensity of poverty in Bamyan (59.4%) is the highest. The fifth column shows that Baghlan has the highest proportion (16.7%) of households that are vulnerable to poverty. The last column indicates that 58.6% of household in Badakhshan suffer from severe poverty; that is the highest rate of severe poverty among the regions.

The above spider graph (Figure 10.5) shows that the poor households are most often deprived in food security, years of schooling, prenatal care, maternal and child health, electricity, sanitation, drinking water, cooking fuel, and source of heating. The graph also reports that most of the poor households own their house and possess some assets – for example, a radio, TV, DVD/VCD/CD/VCR, telephone, bike, sewing machine, tape recorder, or electric fan. Significantly, they also scored well in terms of child school enrolment. Among the SFA regions, the poor households in Badakhshan are more deprived in comparison to the poor households in other SFA programme regions. More specifically, they are more deprived in food security, prenatal care, maternal and child health, electricity, sanitation, cooking fuel, and source of heating.

Figure 10.6 above reports a broken down picture of the regional PAMPIs. The chart displays one column per region: Badakhsan, Baghlan, Bamyan, and Takhar. Different shades inside each bar report the contribution of each weighted indicator to the overall PAMPI value. It reports that food security (FS) and employment (EM) are respectively the highest and second highest contributors to the PAMPI in all the SFA regions.

Table 10.4 Programme Area Multidimensional Poverty Index across programme regions

Region	Programme Area Multidimensional Poverty Index (PAMPI = H×A)	Incidence of poverty (H)	Average intensity across the poor (A)	Percentage of population vulnerable to poverty	Percentage of population in severe poverty
Badakhshan	0.444	75.8%	58.6%	12.7%	58.6%
Baghlan	0.372	64.7%	57.5%	16.7%	45.1%
Bamyan	0.435	73.3%	59.4%	15.5%	54.6%
Takhar	0.423	72.3%	58.6%	14.5%	53.6%

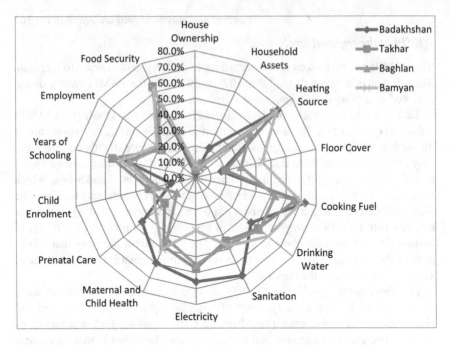

Figure 10.5 Proportion of poor households across different indicators

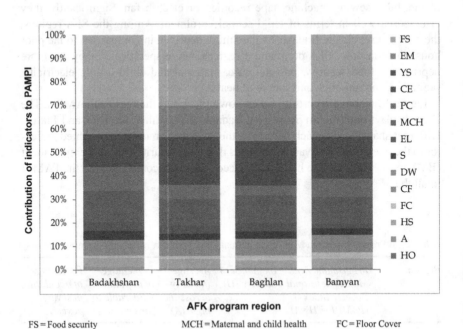

FS = Food security	MCH = Maternal and child health	FC = Floor Cover
EM = Employment	EL = Electricity	HS = Heating sources
YS = Years of schooling	S = Sanitation	A = Asset
CE = Children enrolment	DW = Drinking water	HO = House ownership
PC = Prenatal care	CF = Cooking fuel	

Figure 10.6 Percentage contribution to PAMPI

Use of PAMPI at SFA

The PAMPI will be used as an evaluation tool for the SFA programme regions. The first PAMPI is measured on the basis of the PAS 2014 and disaggregated by the programme regions of Badakhshan, Baghlan, Bamyan, and Takhar. The SFA will calculate a new index based on the second round of the PAS, which is to be conducted in 2019. This will allow the SFA to compare Multidimensional Poverty Index results over time against the aid they have received from the donors. For example, the overall PAMPI 2014 value is 0.42. Hypothetically, if the PAMPI value 2019 is found to be 0.32, the SFA may conclude that the data suggests the poverty situation in SFA programme areas has improved by 24%, as per the equation $(0.42–0.32)/0.42 \sim 0.238$, or 24%. Similarly, the 2019 value (0.32 in this example) will also be broken down according to the SFA programme regions to check regional changes to the indicators in order to ascertain the relative contribution of each indicator to the overall PAMPI 2019 value. The SFA will also calculate the aid they spent for this much improvement in poverty reduction and will develop a strategy to increase aid efficiency. The availability of this sort of poverty lens will provide SFA decision-makers with both a long-term impact assessment tool and an indication of future programming needs.

Conclusion

The PAMPI is an innovative tool for SFA to use to measure the long-term outcome of SFA programmes in poverty reduction and to measure and increase aid efficiency. While there are many tools to evaluate the effect of development interventions, the PAMPI offers the simplest way to measure the impact in poverty reduction over time using only a single and holistic measure that can be calculated from established survey tools.

The PAMPI value is prone to change according to the joint distribution of deprivations in indicators and can be disaggregated depending on specific indicators, regions, ethnicities, and so on. The first PAMPI value (0.42) indicates that approximately 72% of households in SFA programme areas are multidimensionally poor, with an average poverty intensity rate of 58.4%. The PAMPI also clearly indicates that in order to reduce overall poverty, the dimensions of most pressing concern are food security, employment, and health.

The Multidimensional Poverty Index provides a useful analytical lens for measuring poverty, but it is not intended to be used in isolation; rather the PAMPI proposed here is intended as a complementary addition to a broader analytical toolkit – calculating a PAMPI does not prevent senior managers from looking at other indicators of poverty provided by existing survey tools and reporting methods. The index proposed here will provide decision-makers with an additional layer of data to help them corroborate and fine-tune policy choices.

References

Alkire, S. (2010) *The Multidimensional Poverty Index & the Millennium Development Goals: Showing Interconnections*, Oxford, UK, OPHI. Available at: www.ophi.org.uk/wp-content/uploads/OPHI-MPI-MDG-Brief.pdf.

Alkire, S., & Santos, M. E. (2010) Acute multidimensional poverty: A new index for developing countries, *UNDP Human Development Research Paper 2010/11*. Available at: http://hdr.undp.org/en/reports/global/hdr2010/papers/HDRP_2010_11.pdf

Colombia. (2019) *Metas Trazadoras*. Bogota, Republic of Colombia. Available at: www.ods.gov.co/goals/1

CSO. (2014) *Estimated Population of Afghanistan*. Islamabad, CSO. Available at: http://cso.gov.af/en/page/demography-and-socile-statistics/demograph-statistics/3897111

Mitra, S., Jones, K., et al. (2011) Implementing a multidimensional poverty measure using mixed methods and a participatory framework, *Social Indicators Research*, 110(3), pp. 1061–1081. Available at: https://link-springer-com.proxy.bib.uottawa.ca/article/10.1007/s11205-011-9972-9

MPPN. (2019) *The Multidimensional Poverty Peer Network: Connecting Policymakers Globally*. Oxford, MPPN. Available at: www.mppn.org/wpcontent/uploads/2017/03/MPPN_2016_English_A4_5mm_Nov_printers.pdf

National Statistics and Information Authority. (2019) *Afghanistan Multidimensional Poverty Index 2016–2017*. Kabul, NSIA. ISBN: 978-9936-1-0309-2

OPHI. (2019a) *The Alkire-Foster Method in Research and Policy*. Oxford, OPHI. Available at: https://ophi.org.uk/research/multidimensional-poverty/research-applications/

OPHI. (2019b) *Global Multidimensional Poverty Index*. Oxford, OPHI. Available at: https://ophi.org.uk/multidimensional-poverty-index/

Ravallion, M. (2011) On multidimensional indices of poverty, *The Journal of Economic Inequality*, 9(20), pp. 235–248. Available at: https://link.springer.com/article/10.1007%2Fs10888-011-9173-4

Stiglitz, J. E., Sen, A., & Fitoussi, J. P. (2009) *Report by the Commission on the Measurement of Economic Performance and Social Progress*. Paris, Office of the President, Republic of France. Available at: https://ec.europa.eu/eurostat/documents/118025/118123/Fitoussi+Commission+report

UNDP (2018) *Multidimensional Poverty Index (MPI) Human Development Report 2018*. New York, UNDP. Available at: http://hdr.undp.org/en/2018-MPI

Part V

Non-zero options for local development

11 Development aid and access to water and sanitation in sub-Saharan Africa

Sanni Yaya, Ogochukwu Udenigwe, and Helena Yeboah

The sub-Saharan Africa region currently has 32% of its population without access to improved water sources and 695 million people without improved sanitation facilities. Achieving the sustainable development goal of ensuring access to water and sanitation for all by 2030 remains a challenge. Due to financial constraints in their efforts to increase water and sanitation access, governments across sub-Saharan African have been recipients of official development assistance. This chapter discusses the performance of official development aid in sub-Saharan Africa's water and sanitation sector and also reviews factors that determined successes or failures in improving the sector. This chapter also provides updated information on allocated development assistance to the water and sanitation sector in sub-Saharan Africa and the corresponding outcomes. Using the case of the Democratic Republic of Congo, this chapter uncovers factors that impede aid's effectiveness in improving access to a water supply and sanitation. The chapter also discusses the outcomes of different water supply models across sub-Saharan Africa and their impact on access to water and sanitation in the region.

Keywords

Water and sanitation, sub-Saharan Africa, official development assistance, aid effectiveness Sustainable Development Goals, Democratic Republic of Congo

Introduction

Access to essential social services is vital in sustaining human development and achieving development goals. Services such as water and sanitation go beyond the obvious advantage of providing hydration, to saving lives, preventing diseases, and sustaining human health. However, the World Health Organization (WHO) estimated that 844 million people lack access to basic drinking water (World Health Organization, 2018). WHO also estimates that 2.3 billion people do not have access to basic sanitation facilities (WHO, 2018). Addressing water risk factors could have prevented 361,000 child deaths in 2015 and 280,000 annual diarrhoeal deaths due to poor sanitation (WHO, 2018). However, the water and sanitation sector does not receive adequate attention from governments and bilateral and multilateral donors. This is evident as smaller amounts of official development aid (ODA) are made out to the water and sanitation sector compared to other social services such as education, health, and population planning (Salami et al., 2014); this in turn reflects the uneven progress made towards achieving developmental goals particularly in sub-Saharan African countries (SSA) (Winpenny et al., 2016).

Africa faces challenges in access to water. Water availability in Africa is affected by the disparity of its resource endowment across countries. Northern and Southern Africa have some of the driest countries on the continent, while countries in central Africa are endowed with fresh water resources and account for 48% of Africa's water resources (WHO & UN, 2014). This endowment notwithstanding, 25 African countries face the possibility of water scarcity by 2025. A significant global initiative aimed at improving access to water and sanitation was the Millennium Development Goals (MDGs). A key target of the MDGs (Goal 7) was to halve the proportion of the global population without sustainable access to clean and safe drinking water and basic sanitation by 2015. The goal measured progress in access to safe sources of water and sanitation using data from 1990 as baseline. The SSA region failed to meet the MDG targets with 32% of the population not having access to improved water sources and 695 million people using unimproved sanitation facilities (WHO & UNICEF, 2017). Basic water service is one that is free from contamination and within a round trip of 30 minutes. Improved water sources take into account piped water on premises, boreholes, public taps, and rainwater collected and stored in taps. Unimproved sources can include an unprotected dug well, tanker trucks, and surface water. While it is notable that the region achieved a 20% increase in the access of improved water sources since 1990, nearly half of all people still using unimproved water sources as at 2015 resided in SSA (United Nations, 2015). Safely managed drinking water – that is, water free from contamination and available when needed – was accessed by only 24% of the population in SSA, and safely managed sanitation services – excreta safely disposed of in situ – was accessed by only 28% of the population (WHO & UNICEF, 2017).

The move from the MDGs era to the era of Sustainable Development Goals (SDGs) saw a stronger focus on reducing inequalities, not only between regions and countries but within them. This was in response to the stark disparities in access to water and sanitation between urban and rural areas and wealth quintiles within individual countries. For example, at national levels, the majority of Angola's population had access to basic drinking water services in 2015; however, only 20% of the rural population has access, compared to 60% of the urban population (WHO & UNICEF, 2017). Even within urban areas in SSA, cities and peri-urban areas are experiencing population growth and migration from rural areas; this has put immense pressure on the water infrastructure and led to a deterioration of coverage rates. Currently, eight of the world's ten most rapidly urbanizing countries are in SSA, and most of the growth is happening in informal settlements inhabited by the poor (Dos Santos et al., 2017). Access to water and sanitation in these areas is worse than in core urban areas. Therefore, access to water and sanitation in urban areas remained stagnant during the MDG era (Dos Santos et al., 2017). In confirmation, Figure 11.2 below shows that access to pipe water in the urban areas in SSA declined from its value of 66.8% in 2000 to 56.2% in 2015, while access to available water when needed declined slightly between 2000–2015. Access to sewer connections and latrines as shown in Figure 11.4 declined as well.

The gaps between access to water supplies and sanitation in rural and urban populations in SSA remains evident. Only 17.3% of the rural population had access to piped water in 2015, compared to 56.2% access to the urban population. And only 9.5% of the urban population had access to septic tanks in 2015, while a mere 1.7% of the rural population had access (WHO/UNICEF (JMP), 2017).

Figures 11.1–11.4 show the disparities between populations' access to the various service levels of water and sanitation in rural and urban areas.

The impact of inadequate access to water and sanitation are far-reaching and have consequences on attaining development goals. The burden of collecting water often falls on women and girls, and this can have implications for achieving SDG 5, which aims to achieve gender equality and empower women and girls (El Khanji, 2018). Limited access to water threatens girls' school attendance in a region where girls remain at a disadvantage in accessing primary and secondary education compared to boys (UNESCO, 2015). Moreover, it is estimated that women can spend more than six hours a day collecting water from sources that are 30 minutes or more away. The cost of time and labour that women spend in

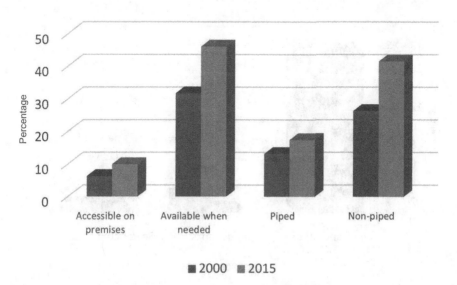

Figure 11.1 Proportion of rural population with access to improved water supply in SSA

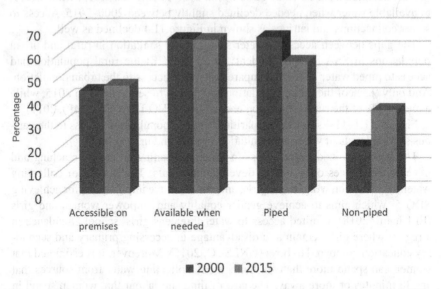

Figure 11.2 Proportion of urban population with access to improved water supply in SSA

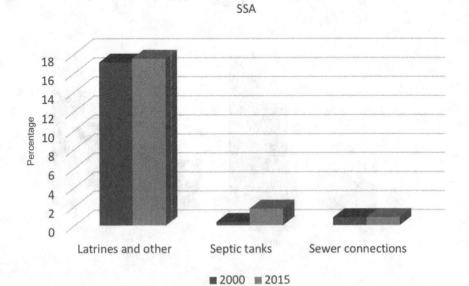

Figure 11.3 Proportion of rural population with access to improved sanitation in SSA

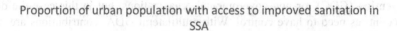

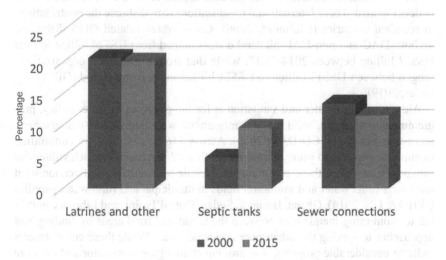

Figure 11.4 Proportion of urban population with access to improved sanitation in SSA

Source: (WHO/UNICEF(JMP), 2017)

collecting water was equated to the average wage of an unskilled labourer in Kenya (El Khanji, 2018). This underscores the need for urgent attention to the water and sanitation sector, particularly when countries in SSA are currently off track in ensuring universal access to water and sanitation by 2030 (Schmidt-Traub & Sachs, 2015). There is a consensus on the importance of bulk investments in water and sanitation in order to achieve the SDGs. It is recommended that these investments go beyond providing access but also sustain efficient water supply and sanitation infrastructures.

ODA investments in water and sanitation in SSA

"Official development assistance" (ODA), a term originating from the Development Assistance Committee (DAC), part of the Organization for Economic Co-operation and Development (OECD), refers to financial grants and loans given to a developing country's official sector with the main purpose of promoting its economic development and welfare. The loans are concessional in nature with a grant element of at least 25%. These forms of assistance can be bilateral or multilateral. Bilateral ODA is sent from donor countries to recipient countries either directly or through international organizations such as the United States Agency for International Development (USAID), the (British) Department for

International Development (DFID), and the Center for Disease Control and Prevention (CDC), or non-governmental organizations, but in this case the donor countries need to have control. With multilateral ODA, contributions are made by OECD member countries to international organizations that specifically work in development. These international organizations then disburse the contributions to recipient countries (Clermont, 2006). Gross water–related ODA disbursed by both DAC and non-DAC bilateral donors ranged from US$24 billion to over US$27 billion between 2014–2017, while that from multilateral organizations ranged between US$17 billion to US$21 billion in the same period (OECD statistics, 2019).

Aid allocated to water and sanitation is for the purposes of water policy, programme and planning, solid waste management, water legislation, management, and water policy (OECD-DAC, 2013). Across Africa, countries are committing to improving access to water and sanitation through strategies to reach vulnerable groups. Examples of these commitments include Mozambique's allocation of at least 40% of its water and sanitation funds to municipal and rural water supplies (WHO & UN, 2014). Ghana, Burundi, Sudan, Cote d'Ivoire, and Liberia committed to monitoring inequalities between the urban and rural areas by finding best approaches to serving the urban poor and rural areas. While these commitments indicate considerable progress, it is proving challenging to monitor and measure progress in reaching vulnerable groups across the continent. WHO reports that less than 40% of countries across Africa monitor water and sanitation service provisions to the poor (WHO & UN, 2014). Specifically, only 33% of countries have policies that include populations living in poverty and 13% monitor the progress made in reaching these vulnerable groups. Furthermore, less than 15% of African countries apply financial measures towards reducing inequalities in access to water and sanitation (WHO & UN, 2014). This has been explained by limited financial capacities of governments in SSA to enhance their water and sanitation sector without international aid.

The importance of aid to the water and sanitation sector has long been recognized. From 1990 to 2006, ODA accounted for 84.4% of aid to SSA. In 2005, donors at the G8 summit committed to doubling aid to Africa to improve delivery of public services and infrastructure in the health, education, and water, and sanitation sectors. Despite this increase, aid provided to the water and sanitation sector increased from 2.8% in 2002 to only 4.1% in 2008 of overall ODA to SSA (Salami et al., 2014). More recent data indicate an increase of 7.6% of total ODA to Africa in 2012. From 2002 to 2017, gross ODA disbursements to all sectors in 50 SSA countries was almost US$360 billion, while only about US$11 billion was directed towards water and sanitation. However, while the level of aid available to the water and sanitation has increased over time, the allocation to the sector is still considered low and a small fraction of the total. Increasing the volume of aid without targeting the water and sanitation sectors may impact on access to the population. Therefore, the next section reviews the literature on the linkages between the availability of ODA and intended outcomes in the water and sanitation sector.

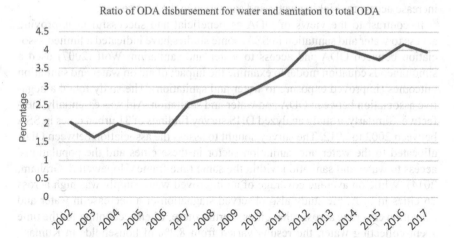

Figure 11.5 Ratio of ODA disbursements for water and sanitation to total ODA

Source: OECD statistics (2019)

Impacts of ODA in SSA

Discourses around official aid to SSA indicate that it is beneficial and success-ful in enhancing economic growth, reducing extreme poverty, increasing access to services, and strengthening social institutions. Addison and colleagues (2005) demonstrated the importance of foreign aid towards SSA in their statistical analysis showing that official aid to SSA between 1960 to 2002 enhanced the standard of living in SSA. They concluded that official aid was essential in achieving the MDGs (Addison et al., 2005). Jeffery Sachs supports this viewpoint in indicating the significant gains made by the MDGs towards improving access to water and sanitation in West African countries. He further urges rich countries to dedicate 0.7% of their gross national product to official development assistance, particularly to SSA (Sachs, 2005, 2014). A report from the OECD indicates that in 2010–2011, US$7.6 billion was committed to water and sanitation with the largest bilateral pro-viders being Japan, Germany, and the United States. Sub-Saharan Africa received 25% of the total aid to the water and sanitation sector. The report indicated that this aid was instrumental in encouraging progress towards water and sanitation MDG targets (OECD-DAC, 2013). A study examined the impact of official development aid to water and sanitation projects in Africa and affirms that aid has helped to improve progress towards gender equity by reducing the time women and girls spent to collect water (Ndikumana & Pickbourn, 2017). Furthermore, the study affirmed that increased access to water and sanitation in SSA helped to improve health and human outcomes, thus improving the overall standard of living. Similar to Sachs, Ndikumana and Pickbourn (2017) assert that funding remains insuffi-cient towards the water and sanitation sector. They emphasized the importance of aid to water and sanitation and the need for influx of official development aid to increase access to drinking water and sanitation in the SSA region.

In contrast to the views of ODA as beneficial and successful in improving access to water and sanitation in SSA, some studies have indicated a limited association between ODA and access to water and sanitation. Wolf (2007) used a simultaneous equation model to examine the impact of aid on water and sanitation outcomes (improved exposure to water and sanitation). The study found a negative association between ODA and water and sanitation. Aid was essentially ineffective. Similarly, a study analyzed DHS survey findings of 31 urban cities in SSA between 2002 to 2012. The survey sought to assess the association between ODA allocated to the water and sanitation sector in these cities and the populations' access to water and sanitation within the same time frame (Hopewell & Graham, 2014). While on average coverage of an improved water supply was high across 26 cities in SSA, the study also observed stagnation or a decrease in water and sanitation coverage across other cities. Furthermore, when accounting for the time spent collecting water, the results varied from 8.2% of households in Kumasi, Ghana, that spent 30 minutes or more collecting water, to 43.3% of households in Yaunde, Cameroon. Improved sanitation coverage was generally low across the 23 cities in SSA (Hopewell & Graham, 2014).

In order to understand the conflicting outcomes of ODA and access to water and sanitation in SSA, studies have examined different determinants for access to water across SSA. While it might be logical to expect that countries with lower levels of water coverage would receive more aid from donors, studies have shown the opposite situation. A study examining global ODA commitments over two decades (1990–2010) towards all developing countries found inequality among developing countries by economic level. About US$59.82 billion was committed to the water and sanitation sector in all developing countries. Overall, ODA for water and sanitation increased after 2002, with an annual addition of US$4.03 million. Countries with the highest need (measured by the mortality rates of children younger than 5 years old due to pneumonia and diarrhoea) were mostly in SSA. However, the largest amount of ODA was committed to India (9.6%), Iraq (9.5%), and Vietnam (7.54%). Countries with the highest needs for water and sanitation were poorly targeted and received less ODA (Cha et al., 2017). Most importantly, water and sanitation projects in SSA have different levels of investments.

These findings were corroborated by Bain and colleagues (2013), who contend that allocation of aid to water and sanitation is not always commensurate with need. Their analysis found that between 2000 and 2010, several countries, which included countries in SSA, with low levels of improved water coverage received low levels of ODA compared to countries with higher coverage (Bain et al., 2013). They observed a cycle whereby successful investments which manifested in higher water coverage in countries led to more donor financing. This could provide explanations for the slow progress observed in SSA despite the presence of ODA. Furthermore, expenditures to the water and sanitation sector were substantially higher in urban areas of countries, even though the rural population in most SSA countries outnumbered the urban population. Studies have offered explanations for the need for more ODA towards urban areas, citing more complex demographics and water systems in urban areas compared to rural areas (Adams et al., 2019).

Another factor that impedes the effectiveness of ODA in SSA is the gap between commitment to ODA and disbursement of ODA, which remains huge. According to Clermont (2006), when donors commit to contribute a certain amount as aid to water and sanitation in developing countries, the amount is included in the total sum of ODA commitments in that year. However, donors can put off the ODA payment for more than ten years. As shown in Figure 11.6, donors made an estimated commitment of about US$407 million to water and sanitation in 2002; however, the total amount disbursed was less than 25%. The year 2008 recorded the lowest percentage of disbursement as a share of commitment, while 2011 recorded the highest (49.4%).

The allocation of ODA to the various water and sanitation projects is also considered important. According to Salami and colleagues (2014), investment in education and training in water supplies and sanitation play a fundamental role in the success of water and sanitation processes as well as an essential part for sustainable development. They, however, found that from 2002 to 2008, large systems of water supply and sanitation had a greater gross ODA disbursement of 50%, followed by 18% to a basic drinking water supply and basic sanitation. Only 3% was allocated to waste management, while education and training in water supply and sanitation had 0.2%, which was the least. According to the OECD, the gross ODA disbursement to water and sanitation in SSA over the period of

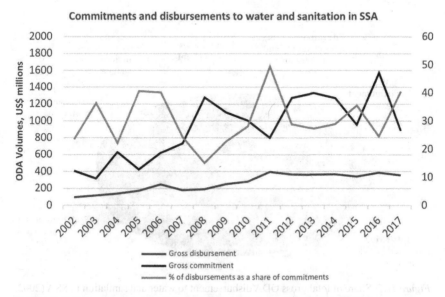

Figure 11.6 Commitments and disbursements to water and sanitation in SSA

Source: OECD statistics (2019)

2002–2017 as indicated in Figure 11.7 had a similar trend except that the basic drinking water supply and sanitation had higher allocation than the water supply and sanitation for large systems. Basic drinking water and sanitation had 46% of the total disbursement, and disbursement for large water and sanitation systems was 30%, followed by water resources policy and administrative management with 15% (OECD statistics, 2019). River basin development, waste management, education and training in water supplies and sanitation, and water resources protection received less than 5% of ODA disbursement each. As economies grow, new industries emerge to respond to the demand and expectations of consumers and businesses. This increases pollutants in drinking water and reservoirs in many cities, which need better treatment technologies (OECD-DAC, 2011). However, most developing countries are not able to rise up to this task as less aid (2%) is allocated to protecting these water bodies as well as that allocated to manage industrial waste (2%).

Furthermore, the level of inequality in service levels between rural and urban populations makes it difficult to achieve greater access to improved water and sanitation, particularly in the rural areas. Figure 11.9 shows the different service levels of water and sanitation services in rural and urban areas. Unsurprisingly, almost 160 million members of the population in rural areas used unimproved drinking water in 2015, compared to 24 million in urban areas the same year. The population in rural areas that engaged in open defecation increased by 5 million between 2000 and 2015. The total unimproved sanitation for the rural areas as of 2015 was over 300% higher compared to urban areas. Open defecation in the urban areas in SSA increased exponentially between 2000 and 2015. According to

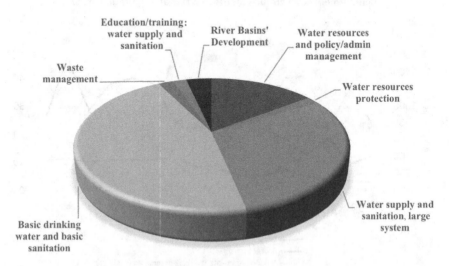

Figure 11.7 Share of total gross ODA disbursement to water and sanitation in SSA (2002–
                2017) by project typology

Source: (OECD, 2019)

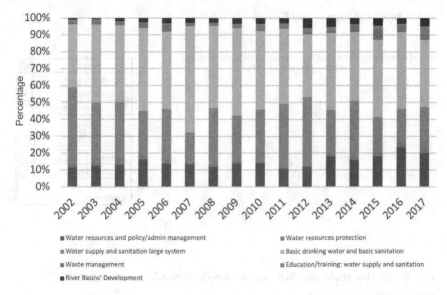

Figure 11.8 Trend of gross ODA disbursement to water and sanitation in SSA over the period of 2002–2017, by project typology

Source: (OECD, 2019)

Table 11.1 Categories of ODA allocation to water and sanitation

Inter-sectoral categories of allocation of ODA to water and sanitation	Definitions
Water resources policy and administrative management	Water sector policy and governance; institution capacity-building and advice; water supply studies and assessments
Water resources protection	Conservation and rehabilitation of surface water; protection of surface water; prevention of water contamination
Water supply and sanitation, large system	Water treatment plants; water storage; pumping stations; conveyance and distribution systems; sewage management; domestic and industrial wastewater treatment plants
Basic drinking water and basic sanitation	Hand pumps, rainwater collection, and fog harvesting; storage tanks, small distribution systems, latrines, on-site disposal, and alternate sanitation systems
Waste management	Municipal and industrial solid waste management, such as toxic waste; collection, disposal, and treatment; landfill areas; composting and use
Education/training: water supply and sanitation	Education and training for service providers and sector professionals
River basin development	Integrated river basin projects, dams, and reservoirs; river flow control

Source: (OECD-DAC, 2019)

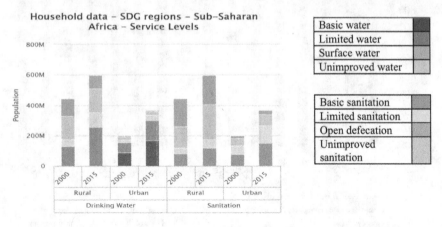

Figure 11.9 Household data, SDG regions, sub–Saharan Africa, service levels

Source: (WHO/UNICEF (JMP), 2017)

Salami and colleagues (2014), the increased population in urban areas in all countries has resulted in unstable access to improved sanitation and moreover, the sanitation sector is oftentimes neglected or given less priority in government's budget allocation. SSA has the highest prevalence of urban slums and the population in these areas is expected to reach 400 million by the year 2020. Rapid migration and unplanned urban centres have led to overcrowding in unstable settlements, and sanitation and water access issues, which have huge consequences on these populations (UNDESA, 2014).

In addition, absorptive capacity of funds also highlighted disparities in aid received and progress made in the water and sanitation sector of countries. Absorptive capacity is affected by a region's institutional capacity – that is, governance of policy and planning on the use of funds (Gomez et al., 2019). The limited ability to monitor or evaluate outcomes of ODA deters ODA commitment from donors. Findings show that at a global scale, countries with weak governance, measured by political stability, control of corruption, and accountability, had lower access to improved sources of water and sanitation. This was evident in sub-Saharan African countries, which had the least access to water (Gomez et al., 2019).

The case of the Democratic Republic of Congo

The Democratic Republic of Congo (DRC), a country in the central part of Africa, is the second largest country in SSA. The country has experienced a series of political and social instabilities in its early years up until 2002, when the Pretoria Accord was signed to establish unity among warring parties (Central Intelligence Agency (CIA), 2018). However, political instability and armed violence is still experienced in DRC. It has been established that since water projects are long-term projects, political stability is a necessary condition for increasing a country's capability to take up aid and also to develop projects (Clermont, 2006). The DRC is endowed with natural resources but remains poor due to corruption, instability, and conflicts, as well as high HIV/AIDS prevalence rates resulting in higher rates of poverty and mortality, and. moreover, the country failed to achieve the MDGs by 2015. The DRC, in addition, has abundant water resources and has internal renewable water supply resources, which are ranked the largest in Africa, but most of the population, particularly the rural population, in the DRC have little access to safe drinking water (USAID, 2010). The DRC is among the least developed countries in SSA and received little commitment of ODA for water and sanitation per person. As at 2017, only 5.3% of total ODA disbursed was allocated to water and sanitation.

Several efforts have been made by international agencies and the government to build an accountable state, promote peace, and increase access to services for the poor, and to reduce violent conflicts (DFID, 2008). The DFID and other major donors, who provide 85% of DRC's ODA, developed a Common Assistance Framework to support the above objectives and the government's action plan for 2008 to 2010, which prioritized six areas for action and five secondary funding areas. It was envisioned that DFID would increase total support for DRC from £70 million in 2008–2009 up to £130 in 2010–2011. Water and sanitation were, however, not part of the six priority programmes, and therefore did not receive much funding and staff compared to some key areas like education, security, roads, and humanitarian assistant (DFID, 2008). By funding the rural water and sanitation programme, the Programme Villages et Ecoles Assainis project and community managed water schemes in Mbuji Mayi; the main outcomes for the water and sanitation plan were to provide sustainable safe drinking water and improved sanitation for approximately 4 million people and strengthen the government's structure in the rural water and sanitation sector, and its capacity to support the provision of safe water (DFID, 2008). However, looking at Figures 11.10 and 11.11 below, there are still issues with access to improved water and sanitation particularly in the rural areas in DRC.

Figure 11.10 shows the trend of water and sanitation service levels among DRC households. Unimproved drinking water consistently increased within the period of 2000 to 2015. While basic drinking water access increased in the rural areas in DRC, it was proportionately lower than that of the urban areas. Over 23 million members of the rural population used unimproved drinking water compared to over 4 million members of the urban populace. In 2000, a little over 15 million members of the rural population used unimproved sanitation, but as of 2015, the number had jumped to almost 23 million. Access to basic sanitation slightly increased within this period, while open defecation saw a continuous increase. Comparing 2000 to 2015, the number of populations that engaged in open defecation almost doubled in the rural areas and almost tripled in the urban areas. The DRC still experiences huge disparities between rural and urban access to

180   *Yaya, Udenigwe, and Yeboah*

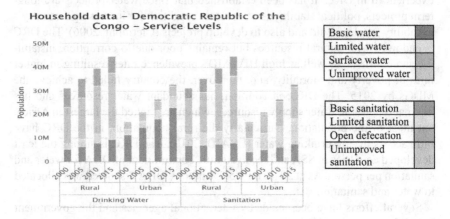

Figure 11.10 Household data, Democratic Republic of the Congo, service levels
Source: (WHO/UNICEF (JMP), 2017)

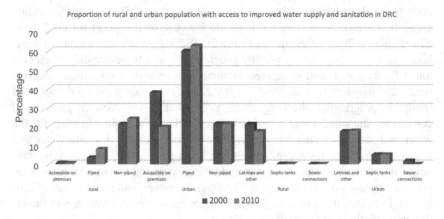

Figure 11.11 Proportion of rural and urban population with access to improved water supply and sanitation in the DRC

Source: (WHO/UNICEF (JMP), 2017)

improved water and sanitation, as seen in Figure 11.11. About 62.7% of the urban population had access to piped water in 2015, compared to only 7.8% access for the rural population. Although access to water on premises available to the urban population nearly halved between 2000 and 2015, only 0.5% of the rural population have access to water on premises. Access to latrines in the rural areas fell in 2015, while that of the urban areas did not see any improvement between 2000 and 2015. Septic tank access for both rural and urban populations remained stagnant between 2000 and 2015.

Water supply models across SSA and outcomes

ODA allocations function not only to promote sustainable development but also to enhance business development and expand the private sector (Jakupec & Kelly, 2016). Privatization of water resources has characterized some time periods in SSA, so have public water supply models. Therefore, it is worthwhile to understand the outcomes of service provisions in SSA particularly in cases when conditionalities are applied to ODA in preference for a specific model – for instance, privatization. This significantly impacts the social conditions within beneficiary countries. Conditionality is used by donors as a means to enhance the likelihood of repayment of aid; sometimes it can be used to ensure that funds are not used inconsistently with donor's values. While conditional aid is beyond the scope of this chapter, its impact has been felt in water supply models in SSA countries such as Ghana. It is therefore important to examine different water supply models and their effectiveness in expanding access to water in SSA.

The decade 1981–1990 ushered in the United Nation's international drinking water supply and sanitation decade with the goal of making water available for all, to promote health (UN, 2019). This came in response to the spread of water-related illness in SSA. In the 1980s, most countries were experiencing severe financial crises and sought financial assistance from donors. However, when they were unable to honour their financial obligations, donors such as the World Bank and IMF instituted structural adjustment programmes whereby cuts were applied to government spending on the water supply and other social services, and privatization of these services was favoured by donors (Adams et al., 2019). An example is in Ghana, where water privatization was explicitly tied to ODA (CCOPD, 2005). Primary objectives for aid were to redefine the role of the state in the provision of public services. Water was seen as an economic good and privatization of water utilities ensued. The first step to privatization in Ghana was to separate non-revenue-generating rural operations from a potentially lucrative urban operation. Water tariffs were then increased in Ghana to recover costs because privatizing water requires a large amount of infrastructure at a high cost. Privatization of water and sanitation services was advocated for by donors in the hopes that competition among different water service providers would generate revenue and expand service delivery (Schmidt-Traub & Sachs, 2015). There is limited evidence of the successes of privatization. The legitimacy of water privatization is questioned based on assumptions that it is the only option to improve efficiency and also improve the conditions for water access to the poor. Water privatizations in SSA were set in effect without civil society consultations thereby de-legitimizing the participatory process. In the case of Ghana, civil society groups opposed the state and donors such as the World Bank and asked for more transparency in the privatization process. Such movements indicate the staunch resistance to water privatization even though privatization still operates in certain countries such as Cote d'Ivoire, Mozambique, Niger, Senegal, and South Africa (Adams et al., 2019; CCOPD, 2005). Privatization of water and sanitization services in SSA is largely considered a failure in improving water access to the population. Failure of privatized water supply systems has been attributed to unsuitable policy environments for private sector participation. Furthermore, given the essential nature of water, there has been public discontent regarding the emphasis on profit from water, which is seen as a public good.

With the failure of privatization, many African governments opted for public-private partnerships, which involve the management of the services by private entities but under state ownership. Private-public partnerships failed to improve water coverage in urban Africa because private corporations picked wealthier neighbourhoods over poor ones (Adams et al., 2019). There was also the issue of political interference in fraudulently awarding contracts. Furthermore, the state of the water system infrastructure was already in a precarious state beyond what the private-public investments could address. Informal water supply models are emerging in SSA whereby communities partner with private operators. This takes different forms such as a water utility provider working with a domestic or community-based private operator to supply and manage water and sanitation infrastructure in their communities (Adams et al., 2019). In Malawi, a system whereby partnerships were formed between a publicly owned water utility and the community has proven beneficial. Compared to the old system, which was managed by private operators, the new system led to an improved water supply to communities in Malawi through stabilizing prices for water and operating with accountability and transparency (Adams et al., 2019). There are also community-based self-provision partnerships, where communities form their own institutions for water delivery without formal partnerships with governments or utility operators. In countries such as Zambia and Tanzania, communities have come together to dig wells and boreholes, and have purchased water pumps and constructed storage tanks to meet their own water needs (Adams et al., 2019). Apart from enabling communities to mobilize on their own to meet their water and sanitation needs, these emerging initiatives are attractive to foreign aid for sustainability. While the World Bank recognizes that water and sanitation are human rights of individuals, it also argues that a human rights framework does not provide a right to free water. Instead, a human rights framework implies that water and sanitation be affordable and available (The World Bank, 2016). The World Bank's current stance on the issue of water supplies recognizes that there is no one solution to addressing the complex water and sanitation challenges in SSA. It opines that current strategies to address access to water and sanitation must be tailored to the local context, and beneficiaries get to decide what model of water and sanitation services is most appropriate for their context (The World Bank, 2016).

Conclusion

This chapter discussed the state of water and sanitation in SSA. The SSA region failed to meet the 2015 MDG targets with 32% of the population not having access to improved water sources and 695 million people using unimproved sanitation facilities. SSA countries are also currently off track in ensuring universal access to water and sanitation by 2030. The increase in the level of aid available to the water and sanitation over time is evident; however, it is still considered low and a small fraction of the total. Additionally, the allocation of aid to countries for water and sanitation is not always commensurate with need; this exacerbates the already challenging state of ODA. Evidence shows that official development aid to water and sanitation projects in Africa has helped to improve access to water and sanitation and improved progress towards gender equity by reducing the time women and girls spend to collect water. On the other hand, studies observed stagnation or a decrease in water and sanitation coverage across cities even with the presence of ODA. Gaps in access to water and sanitation in SSA are evident between urban and rural areas. Within urban areas in SSA, cities and peri-urban areas are experiencing population growth and migration from rural areas; this has put immense pressure on the water infrastructure and led to a deterioration of coverage rates.

For SSA countries to achieve the SDGs by 2030, there should be at least a 100% increase in ODA for water and sanitation by 2020, which should be directed particularly to the most underserved and vulnerable countries like the DRC, where these forms of assistance are needed most. There should be a focus on system strengthening, equity, and sustainability, with increasing transparency and technical assistance. The gap between commitment and disbursement needs to be bridged using alternate financing sources such as government investments and foreign direct investments. Emphasis should also be placed on investments in education and training on effective and vital water and sanitation practices, waste management, and water resources protection, in order to improve water quality and make effective use of the already scarce water supply resources in SSA. Additionally, improving the water supply in SSA will require partnerships involving strong community participation. The World Bank recognizes that there is no one solution to addressing the complex water and sanitation challenges in SSA. It is therefore important that strategies to address access to water supplies and sanitation be tailored to the local context and that beneficiaries make the decision on what model of water and sanitation services is most appropriate for their context.

References

Adams, E. A., Sambu, D., & Smiley, S. L. (2019) Urban water supply in Sub-Saharan Africa: Historical and emerging policies and institutional arrangements, *International Journal of Water Resources Development*, 35(2), pp. 240–263. https://doi.org/10.1080/07900627.2017.1423282

Addison, T., Mavrotas, G., & McGillivray, M. (2005) Aid to Africa: An unfinished agenda. *Journal of International Development*, 17(8), pp. 989–1001. https://doi.org/10.1002/jid.1255

Bain, R., Luyendijk, R., & Bartram, J. (2013) *Universal Access to Drinking Water: The Role of Aid*. World Institute for Development Economic Research. Available at: www.wider.unu.edu/sites/default/files/WP2013-088.pdf [Accessed 18 April 2019].

CCOPD. (2005) *World Bank Conditionality in Water Sector Privatization Cases from Ghana and the Philippines*. Canadian Catholic Organization for Development and Peace (CCODP).

Central Intelligence Agency (CIA). (2018). The world factbook: Congo, democratic republic of the. Available at: www.cia.gov/library/publications/the-world-factbook/geos/cg.html [Accessed 15 April 2019].

Cha, S., Mankadi, P. M., Elhag, M. S., Lee, Y., & Jin, Y. (2017) Trends of improved water and sanitation coverage around the globe between 1990 and 2010: Inequality among countries and performance of official development assistance, *Global Health Action*, 10(1), np. https://doi.org/10.1080/16549716.2017.1327170

Clermont, F. (2006). *Official Development Assistance for Water from 1990 to 2004*. World Water Council.

DFID. (2008) *Democratic Republic of Congo Country Plan*. London, DFID.

Dos Santos, S., Adams, E. A., Neville, G., Wada, Y., de Sherbinin, A., Mullin Bernhardt, E., & Adamo, S. B. (2017) Urban growth and water access in sub-Saharan Africa: Progress, challenges, and emerging research directions, *Science of the Total Environment*. https://doi.org/10.1016/j.scitotenv.2017.06.157

El Khanji, S. (2018) An empirical exploration of relationships between official development assistance (ODA) and advances in water and sanitation subsectors, *Cogent Economics and Finance*, 6(1), pp. 1–19. https://doi.org/10.1080/23322039.2018.1437661

Gomez, M., Perdiguero, J., & Sanz, A. (2019) Socioeconomic factors affecting water access in rural areas of low and middle income countries, *Water*, 11(2), p. 202. https://doi.org/10.3390/w11020202

Hopewell, M. R., & Graham, J. P. (2014) Trends in access to water supply and sanitation in 31 major sub-Saharan African cities: An analysis of DHS data from 2000 to 2012, *BMC Public Health*, 14(1), np. https://doi.org/10.1186/1471-2458-14-208

Jakupec, V., & Kelly, M. (2016) Development aid_ regulatory impact assessment and conditionality, *Impact Assessment and Project Appraisal*, 34(4), pp. 319–329. https://doi.org/10.1080/14615517.2016.1228339

Ndikumana, L., & Pickbourn, L. (2017) The impact of foreign aid allocation on access to social services in sub-Saharan Africa: The case of water and sanitation, *World Development*, 90, pp. 104–114. https://doi.org/10.1016/j.worlddev.2016.09.001

OECD. (2019) OECD statistics. Available at: https://stats.oecd.org/Index.aspx?lang=en&SubSessionId=&themetreeid=3# [Accessed 20 April 2019].

OECD-DAC. (2011) *Water and Development: Taking Lessons from Evaluation*. Berlin, OECD-DAC.

OECD-DAC. (2013) Financing water and sanitation in developing countries: The contribution of external aid. *The Contribution of External Aid*, 8, pp. 1–8, June. Available at: www.oecd.org/dac/stats/Brochure_water_2013.pdf [Accessed 18 April 2019].

OECD-DAC. (2019) Aid to the water and sanitation sector. Available at: www.oecd.org/dac/stats/water-relatedaid.htm [Accessed 20 April 2019].

Sachs, J. (2005) *The End of Poverty*. The Penguin Press. https://doi.org/10.1353/lag.2006.0020

Sachs, J. (2014) From millennium development goals to sustainable development solutions, *Current Science*, 106(4), pp. 495–496. https://doi.org/10.1016/S0140-6736(12)60685-0

Salami, A. O., Stampini, M., Kamara, A. B., Sullivan, C. A., & Namara, R. (2014) Development aid and access to water and sanitation in Sub-Saharan Africa. *Water International*, 39(3), pp. 294–314. https://doi.org/10.1080/02508060.2013.876570

Schmidt-Traub, G., & Sachs, J. D. (2015) Financing sustainable development: Implementing the SDGs through effective investment strategies and partnerships. *Sustainable Development Solutions Network*, pp. 1–155.

UN. (2019) UN documentation: Development. Available at: https://research.un.org/en/docs/dev/1981-1990 [Accessed 15 April 2019].

UNDESA. (2014) International decade for action water for life 2005–2015. Available at: www.un.org/waterforlifedecade/africa.shtml [Accessed 18 April 2019].

UNESCO. (2015) *Education for All 2000–2015: Achievements and Challenges*. EFA Global Monitoring Report.

United Nations. (2015) The millennium development goals report 2015, *United Nations, New York*, 1–73. Available at: https://visit.un.org/millenniumgoals/2008highlevel/pdf/ MDG_Report_2008_Addendum.pdf [Accessed 18 April 2019].

USAID. (2010, March) Democratic Republic of the Congo water and sanitation profile. Retrieved August 14, 2019, from http://www.washplus.org/sites/default/files/drc 2010.pdf

WHO/UNICEF (JMP). (2017) Household data on drinking water and sanitation. Available at: https://washdata.org/data/household#!/ [Accessed 19 April 2019].

WHO. (2018) WHO | Sanitation. Available at: www.who.int/en/news-room/fact-sheets/ detail/sanitation

WHO & UN. (2014) *Investing in Water and Sanitation: Increasing Access, Reducing Inequalities*. Un Glass (Vol. 1). https://doi.org/9789241508087

WHO & UNICEF. (2017) *Progress on Drinking Water, Sanitation and Hygiene*. World Health Organization.

Winpenny, J., Trémolet, S., Cardone, R., Kolker, J., Kingdom, B., & Mountsford, L. (2016) Aid flows to the water sector, *Aid Flows to the Water Sector*, November. https://doi. org/10.1596/25528

Wolf, S. (2007) Does aid improve public service delivery? *Review of World Economics*, 143(4), pp. 650–672.

The World Bank. (2016) FAQ – World Bank group support for water and sanitation solutions. Available at: www.worldbank.org/en/topic/water/brief/working-with-public-private-sectors-to-increase-water-sanitation-access [Accessed 20 April 2019].

World Health Organization. (2018) WHO | Drinking-water fact sheet. Available at: www. who.int/news-room/fact-sheets/detail/drinking-water [Accessed 10 April 2019].

12 From more spending to better spending

The case of food and nutrition security in Ethiopia

Sanni Yaya, Neville Suh, and Richard A Nyiawung

Ethiopia has for several decades been a very fragile state, challenged with a series of intertribal conflicts between several ethnic groups. Ethnic groups such as the Oromo and Somali, according to UNOCHA (2018), continue to have contentions over arable and grazing land, thus making its food security and interventions landscape very complicated. Natural disasters such as prolonged drought have profoundly affected the agroecology of the country, thus hindering its agricultural productivity – productivity that is needed to meet the food and consumption demands for millions of Ethiopians. Poorer households are more impacted by these environmental shocks and the numerous famines the country has gone through since the 1980s. Inappropriate land use and land tenure policies equally contribute negatively to issues of food security as access to resources and property rights limits agricultural investment and technology adoption for those interested. There has been an increase in the number of food insecure households, with a rising rate of more than 3% per annum in the last quarter century in Ethiopia. Many households are dependent on food aid and assistance programmes and receive up to 19% of all food aid in sub-Saharan Africa (SSA) from international donors to help reduce issues of famine and food shortages. However, the country has made provisions devoted to the assessment of its effectiveness to reduce its reliance on donors' food aid and assistance, as well as provide incentives to domestic farmers and food traders to boost production efforts.

Keywords

Ethiopia, food security, food aid and assistance, government policies, famine, and interethnic conflicts

Introduction

Ethiopia is one of the landlocked nations of Africa, and amongst the most populated countries in Africa, with a projected population of more than 139 million people by 2020 (Tegegne & Gebre-Wold, 1998). Located at the horn of Africa, Ethiopia is bordered by Somalia, Eritrea, Sudan, South Sudan, Kenya, and Djibouti. A rapidly growing population directly exerts pressure and stress on the country's available land, thus exacerbating issues of land degradation. Land degradation amongst other environmental stresses such as deforestation, rapid soil depletion, and erosion, as well as poor vegetation cover challenges both crop and livestock production in the country (Taddese, 2001; Grepperud, 1996). Rapid loss in biodiversity and an increase in desertification due to weak or inappropriate land use system and tenure policies in the country (see Grepperud, 1996) equally hinder agricultural production. Limited investments in the agricultural sector and the effects of climate change through severe drought conditions present challenging condition to local livelihood and most importantly food security for the rapidly increasing population (Gray & Mueller, 2012).

Ethiopia is listed amongst many other countries heavily hit by severe famines in the 1970s, 1980s, and beyond. Causalities of these famines included more than 800,000 lives being lost, with many Ethiopians lossing their household and personal assets (Dorosh & Rashid, 2013). The country had experienced several episodes of famine in the 1950s, although causalities were decreased due to the increasing availability of public health services in many areas, especially rural areas (Kidane, 1989). However, the great famine in Ethiopia within the late 1990s was attributed to environmental factors (drought specifically), which exacerbate the vulnerability of several types of communities to famine situations, especially for many resource-poor rural households (Mariam, 1986). Webb and colleagues argue that these environmental factors interact directly with policy failures to promote the exposure of households to famine situations (Webb et al., 1994). Some of these policy failures include high rates of unemployment, poor investments in agricultural and technology, low agricultural productivity, poor health conditions, and a poverty-driven country (Webb et al., 1994). Poorer households are more impacted by these environmental stresses and natural disasters, which makes them vulnerable to hunger and other devastating effects.

Aside from the constant exposure of Ethiopia to drought and its severe famine conditions, war and conflicts have been another issue plaguing Ethiopia for decades (Dorosh & Rashid, 2013; Tache & Oba, 2009). The Horn of Africa, especially Ethiopia, has experienced several civil wars and intertribal conflicts; these contribute to worsening the country's food security issues (see Keller, 1992). Interethnic/intertribal conflicts have seriously affected and contributed enormously to raising issues of food security in Ethiopia. The country has a history of weak government policies on land use and tenure systems, which often result in struggles and infighting between grazers and various pastoral groups in the country (Tache & Oba, 2009). Beyond policy failures, political differences and politics have been a major source of interethnic conflicts in the country, especially over how resource boundaries are allocated (Teferi, 2012). As some scholars have noted (e.g., Markakis, 2003; Bassi, 1997), these conflicts and guerrilla wars often result in the displacement of millions of persons, particularly pastoral groups, thus affecting livestock productivity, required to meet the needs of the population. As outlined, the aggravated and persistent drought situations and the continuous vulnerability of Ethiopia to climate change effects play a crucial role in creating famine conditions in Ethiopia. Although many attribute exposures to drought as a causal factor for famine, interethnic conflicts, weak government policies, politics, and the fragility of the country due to conflicts have posed severe challenges to food security and better livelihood for Ethiopians for several decades now. In addition to these internal factors, regional and external factors such as a long-time clash with Eritrea and the River Nile question (underdevelopment and underutilization of shared/cross-border resources in the Nile, etc.) have also contributed to exacerbating food insecurity in the country.

Food security challenges – the Ethiopian context

For several decades the country has been the most food insecure country, plagued by famine and severe drought conditions. Interethnic conflicts, wars, and weak government policies all interact with each other to aggravate the food insecurity issues of the country tremendously, to critical levels (Dorosh & Rashid, 2013; Bogale et al., 2006; Ramakrishna & Demeke, 2002). There has been an increase in the number of people who need food assistance, and in the past two-and-a-half decades, the number of people who are food insecure has increased from 5% (in the 1970s) to more than 20% (in 2003) (Gebreselassie, 2006). The number of food insecure people has been rising at a rate of more than 3% per annum in the last quarter century in Ethiopia (Berhanu, 2003). This has led the country to be highly dependent on food aid intervention programmes to assist its local population in maintaining their livelihood, which is crucial for survival (Gebreselassie, 2006). More than 90% of the livelihood of Ethiopians depends on a low-input, rain-fed agricultural farming system; which hardly meets the consumption demand for its rapidly growing population, thus, raising issues of food insecurity and other challenges (Keulen, 2014; Kaluski et al., 2002). Unarguably, there have been numerous food aid interventions in the country, since the early days of the great famines to the present, with much donor and organizational assistance (Rahmato et al., 2013). However, Ethiopia remains one of the most food insecure countries in the world, with more than half of its population malnourished or hungry (UNICEF, 2017).

Agriculture contributes significantly to Ethiopia's export market earnings and its gross national product. More than 60% of the country's population resides in rural areas and depend directly on agriculture for their livelihood, thus, meeting consumption needs of its people is a significant priority (Worku et al., 2012). Enhancing food security measures in the country is very important; however, vulnerable and food insecure communities do vary across the country, with some being more heavily impacted than others, depending on their access to natural resources (Beyene & Muche, 2010). Mitiku and colleagues (2012) outline some of the challenges to food security in Ethiopia; these include but are not limited to the usual agronomic and environmental challenges (that is, pests and diseases, drought, poor soils, etc.), rapid population growth, and weak government policies. The food insecurity in the region is also a result of erratic rainfall, droughts, degradation of the ecosystem, low productivity of the agricultural sector, inadequate technology employed in the agricultural sector, and poor infrastructure (Endalew et al., 2015). Moreover, the reliance of the country on continuous food aid for several decades, especially during and after the great famines, directly affects food security issues in the country to exacerbating levels; this is coupled with reluctant government policies usually affected by political issues, civil conflicts, and wars (Jayne et al., 2002).

Furthermore, Ethiopia's low – input, rain-fed agriculture is exposed to climate change scenarios which affect its agricultural productivity and potentials (Kassie et al., 2014). Access to land and land tenure for agricultural activities remains a contested issue within communities and for the country's government. No clear land ownership and property rights policies exist in Ethiopia, and has impacted the ability to improve agricultural practices, especially for farmers through feasible policies (Tolossa, 2003). Many of its population have limited access or ownership of land and this directly affects their farming decisions and the interest of many people to be engaged in the sector and meet the country's food demand. Hence many of these households have constant food shortages and are food insecure (Bogale et al., 2006). Gender inequality is another issue that play a significant role in household food insecurity. Gender disparities exist concerning access to land, with women being at a disadvantage (see Adal, 2006). Women constitute close to 50% of all agricultural households in Ethiopia, indicating how important they are in supporting better agricultural productivity and economic development (Adal, 2006). With such inequality between men and women concerning access to land or land ownership, female-headed households are usually more prone to issues of food insecurity in the country.

Although constant exposures to drought, conflicts, and weak government policies contribute directly to an aggravated level of food insecurity in Ethiopia, the country's food system itself is very complicated. The lack of varying food sources and options makes it difficult for Ethiopian families to consume a healthy and balanced diet daily with the necessary essential nutrients and micronutrients (Gebru et al., 2018). Ethiopia has the highest number of people who are undernourished and hungry in Africa, with about 32 million people suffering from hunger and its related health concerns, especially children (UNHCR Ethiopia, 2018; Endalew et al., 2015). Additionally, close to 50% of the country's population, which reside mostly in rural areas, live below the poverty line (Mohamed, 2017). With all these numerous challenges and the vulnerability of the country to climate shocks, rapid population growth, and conflicts, there is a need for intervention programmes in the country to help address issues of food insecurity. In response to the issues of food insecurity and its potential threats, many international organizations together with the Ethiopian government have instituted food aid intervention programmes and food aid policies that promotes and advances agricultural technologies and food production.

Ethiopia's disaster preparedness and food security policies

As a point of reference, "Ethiopia's officially food aid policy critically outlines that no able-bodied person should receive food aid without working on a community project in return, supplemented by targeted free food aid for those who cannot work"

(Quisumbing, 2003, 1). The resurgence of food aid policies in Ethiopia started at the end of the 1991 civil war, and since then, the government has carefully increased its measures to avoid the recurrence of the mortality levels experienced during the 1984 and 1985 great famines (Lind & Jalleta, 2005). Several policy timelines do exist in Ethiopia as regards disaster relieve and food security. The National Policy on Disaster Prevention and Management (NPDPM) was promulgated in 1993, with a focus on addressing the root causes of food shortages and famine and on reducing long-term exposure to disasters (TGE, 1993). In 1995, the Disaster Prevention and Preparedness Commission was created to replace the Relief and Rehabilitation Commission and linked with a shift towards reducing vulnerability and matching relief to development (Devereux, 2000). In 1996, the Food Security Strategy was initiated and revised in 2002 with a focus on shifting from being aid in-kind to cash-based; this was introduced through the Employment-Based Safety Net approach to addressing the problem of severe food insecurity within communities (Lind & Jalleta, 2005). In November 2003, "The New Coalition for Food Security Program in Ethiopia" was launched following the famine and drought of 2002–2003 (Gebreselassie, 2006). In 2013, a new policy was developed which included a framework for Disaster Risk Management (DRM) measures, a modification of the 1993 NPDPM (Ethiopia-Government, 2013). The new policy consists of an early warning and risk assessment, decentralized DRM system, capacity building, information assessment, and a disaster risk reduction plan.

Food aid and intervention programmes in Ethiopia

Severe food insecurity, acute poverty, and hunger are still very alarming and continue to challenge Ethiopians. This has necessitated numerous calls by the government and the international community for food-based intervention programmes in the form of aid and assistance (Jayne et al., 2002). In the past three decades, Ethiopia has been hit with more than five major devastating droughts (Levine et al., 2011; Webb & von Braun, 1994) and is highly dependent on international food aid, receiving up to 19% of all food aid in sub-Saharan Africa (World Food Programme – WFP, 2011). Ethiopia, being a country that is frequently drought-affected, poverty driven and highly dependent on food aid. Ethiopia received a total of about 10 million metric tons of cereal as food aid between 1984 and 1998, approximately 10% of its local yearly cereal production (Jayne et al., 2002). From 1998 to date, a range of between 5 and 14 million people have benefited from food aid programmes yearly (Devereux et al., 2006). However, as Sabates-Wheeler and Devereux (2010) posit, despite this food aid, Ethiopia has not experienced any significant progress in addressing vulnerability and rural poverty, although new programmes present a novel attempt to address food insecurity and break the high dependence on food aid by Ethiopians. More than 5 to 6 million persons are in continuous need of food aid, even in years with good agricultural production. Approximately 8 million people could join the number of food insecure persons if the prevailing factors are not addressed by existing intervention programmes (Gebreselassie, 2006).

Two main intervention programmes have been used in Ethiopia to address food insecurity and reduce poverty (Kamenidou et al., 2017; Nega et al., 2010). Considering the country's poverty levels and resource-use challenges, there is a need to have a long-term support system to address emerging food deficiencies (Azadi et al., 2017). Food assistance has become part of the feature of Ethiopia since the onset of famines in 1973–1974, which led to the starvation of over 1 million people (Gebreselassie, 2006). The famine necessitated the establishment of a food aid programme as an ideal relief strategy, with the emergence of the Food for Work (FFW) programme as a core of development programmes in severe food-insecure communities (Quisumbing, 2003). In the 1980s, a massive national FFW afforestation and soil conservation project was set up and managed by the government using labour brigades (Gebremedhin & Swinton, 2000). With an increasing dependency on food aid relief, the government of Ethiopia and international donors needed to increase accountability for the use of food aid on development activities through the FFW programme (Barrett & Clay, 2001; Sharp, 1998).

Food aid in Ethiopia was also being distributed for free (Del Ninno et al., 2007). Under the free distribution programme, cereals (maize, sorghum, and wheat) were handed over to households, while in the case of the FFW programme, participants were typically engaged in community development projects, such as terrace, infrastructure, road, and dam construction (Quisumbing, 2003). Unfortunately, the FFW programme was not successful because preconditions needed to sustain such a long-term programme weren't established. The land policy which distorted property rights for investments generated tenure uncertainty and made it hard to trace the economic agents that were responsible for the cost of non-investment or the benefit of investment, thus affecting the programme (Gebreselassie, 2006). According to Nega and colleagues (2010) and Segers and colleagues (2008), the FFW programme did not have a significant and strong positive impact on the participants' transitory and chronic poverty levels. The FFW programme was focused on the middle-class and rich elites (Azadi et al., 2017), and, moreover, an effort to link relief activities with development programmes through FFW programmes was not as effective as planned.

According to Gebreselassie (2006), "The New Coalition for Food Security Program in Ethiopia" that emerged in November 2003 following the famine and drought of 2002–2003 was intended to solve two fundamental problems. First, the impact of decades of food assistance shipments into Ethiopia, which had led to a dire situation in food production, marketing, and economic development, further creating dependence on food aid; and second, the fact that food aid donors and the Ethiopian government have failed to address the long-term impact of continuous food assistance on the country's economy and the possible risk to the economy if there were to be a sudden end in food aid. The influx of food assistance can reduce the price of produce through its impact on demand and supply (Sabates-Wheeler & Devereux, 2010). Food assistance into a community can surge supply and as a result, reduce demand. This is the case in most developing countries, where domestic food prices fall when there is an inflow of food aid into the country (Abdulai et al., 2004).

The extent of market distortion is dependent on the targeting of food aid. The supply of food aid in most cases within poorer communities indirectly reduces the production potentials of the local people with the population depending sole on this external support rather than looking for alternative solutions (Abdulai et al., 2004). Devereux (2000) described that there is a disincentive effect from constant food assistance on local food production. The dependency on food aid makes many people not interested in food production anymore even in years with favourable environmental condition s for production. However, the New Coalition for Food Security Program in Ethiopia has not been able to limit the quantity of food aid shipments into Ethiopia to a significant level by base on a system that encourages and supplements local production.

Because of the weaknesses of the FFW programmes, the Productive Safety Net Programme (PNSP) was initiated in 2005 to enable Ethiopia to create a food deficiency support system (Nega et al., 2010). The PSNP in Ethiopia was developed by a joint body of the Ministry of Agriculture and Rural Development and the Food Security Coordination Offices (Sharp et al., 2006; Kay et al., 2006). According to Gebreselassie (2006), the PSNP is a social protection scheme implemented by the Federal Food Security Coordination Bureau for people who are chronically food insecure. The PSNP was developed mainly to assist households and individuals that are experiencing acute and chronic food insecurity, and to assist disabled persons with low purchasing power by helping them with income or income-generating activities (Nigussa & Mberengwa, 2009). Two major projects found under the PSNP included the Public Work (PW) and Direct Support (DS) projects (Gilligan et al., 2008). The first and most essential part of these programmes is sufficient PW to accumulate community assets using labour from chronically food insecure households. The second part is the DS, which focuses on those who are disabled.

Further, the PSNP attributes 80% of its funds to DS programmes through cash transfers and food, while 20% is for the administrative and capital cost to manage PW projects (Gebreselassie, 2006). At the end of a maximum of five consecutive years, beneficiaries are supposed to have graduated out of the programme or be self-sufficient. One key feature of the PSNP is that transfers to victims are in the form of cash, which has as the objective to stop the cycle of reliance on donor food aid and to support incentives for domestic farmers and food traders instead of undermining incentives.

However, the particular point of difficulty of such food aid is the ongoing debate of its importance and effectiveness to sustainably improve the lives of the poorest households through agricultural activities and the pilling up of assets. Critics have ascertained that such programmes have not reduced the chronic food insecurity problems, nor have they prevented the depletion of assets (De Rudder, 2013; Bishop & Hilhorst, 2010; Dercon & Krishnan, 2004). Critics have also identified that the process of implementing food aid programmes in local communities could have an obstructive impact. The changes in food aid distribution across communities (which community should benefit from aid, to what amount, and when) over time signal that there is still the possibility to improve the scope (Dercon & Krishnan, 2004). Nevertheless, De Rudder (2013) found that an inverse and significant relationship exists between the percentage of the people who take part in the PSNP in Ethiopia and the per capital in agricultural production (district level), with a lower percentage of beneficiaries in the PSNP having higher agricultural output per capita, and vice versa. Bishop and Hilhorst (2010) carried out an analysis of progress made in implementing the PSNP in the Amhara region. The authors disputed that the execution of the PSNP changed its purpose and was helping the middle-income group instead of the most susceptible households.

According to the PSNP project memorandum, "The primary targeting objective (of the PSNP) should be to guarantee timely and adequate transfers to the most food-insecure people in the most food-insecure areas" (DFID Ethiopia, 2005). It suggested that the targeting transfers to chronic food insecure households have been successful under the PSNP through the implementation of geographic and community-based targeting (Coll-Black et al., 2012). From findings, there has been progress in the targeting methods of the PSNP and the support provided by the programme has been regarded as one of the best direct target programmes at the national level (Coll-Black et al., 2012). Evidence from households suggests the importance of food aid programmes in decreasing household susceptibility in Ethiopia. Thus, it will be beneficial if the targeting of such programmes is improved (Dercon & Krishnan, 2004).

However, the targeted principles of the PSNP found in the Project Implementation Manual were not strictly followed in some regions. In a study by Nigussa and Mberengwa (2009) in Kuyu Woreda region for example, the authors realized from the evaluation that the process of selection of beneficiary participants had been affected by huge discrimination from the PSNP implementation body: the authors found that the selection process was tainted by nepotism, tribalism, and corruption. Further findings from the survey reveal that, as concerns the perception of households as a major criteria for selecting targeted beneficiaries; 53.2% of participant confirmed that the programme targeted the poor; other key factors included; the political position of the household heads; the family size and the number of elderly and disabled persons in a household were taken into consideration.

In another survey in southeast Ethiopia by Welteji and colleagues (2017), beneficiaries raised concerns regarding the management of PW being weak, and spending an inadequate amount of attention on the needs and members of communities in initiating, selecting, and carrying out construction on farm plots and in areas exposed to animal damages. This has led to the unsustainable management of projects by beneficiaries, and there is inadequate monitoring from the PSNP management bureau as most communities prefer to focus on activities that will give them immediate benefits. It is also noted that there is the absence of clear and functional institutional mechanisms that carefully select food insecure people who ought to be included and benefit from the programme (Nigussa & Mberengwa, 2009). This is due to the high levels of administrative and bureaucratic bottlenecks in the PSNP framework, as well as the lack of awareness and low educational levels of the targeted food insecure households. The mode of transfer to beneficiaries is either in cash payments or food assistance and most of the affected households complain of the fact that supplies are often not delivered on time; with some it takes up to a year to receive supplies or payments. Beneficiaries needs are not met in their various villages, and some have to travel from villages to towns and are forced to stay up for to two days to wait for the payment of the transfers and consequently incur the costs on accommodation, meals, and transportation. The PSNP has no evidence of reducing acute or chronic undernutrition in most food insecure communities that have poor diets for children and mothers who have no contact with health agents to get information on good eating habits (Berhane et al., 2017). Concerns have also been raised that the segmented nature and the timing of the transfer of funds hinder the massive purchase of food items during peak periods after harvesting when supply exceeds the demand and prices have fallen; this results in purchases during periods of severe food shortages (Nigussa & Mberengwa, 2009). Kay et al. (2006) found out that some non-beneficiaries who were chronically food insecure thought that their exclusion from the PSNP was due to a lack of relatives or friends among the community decision-makers. Also, landlessness was used as a factor to exclude people from the PSNP, which is inappropriate as there are no clear land tenure policies in Ethiopia, with women household heads being the most disadvantaged group.

Trend in cereal food aid from the WFP and domestic cereal production in Ethiopia

Figure 12.1 shows that cereal food aid shipments and cereal production in Ethiopia between the year 1993 to 2015 has been fluctuating with a slight decrease in food aid and a sharp increase in local cereal production. Figure 12.2 shows that cereal food aid shipments as a percentage of domestic cereal production in Ethiopia has been decreasing from 1993 to 2015. This indicates that, with the numerous intervention programmes, the contribution of domestic cereal production to food security has been increasing even though cereal production has not yet been able to meet local consumption demands. This is because there has been an increase in the number of food-insecure people from 5% in the 1970s to more than 20% in 2003 (Gebreselassie, 2006); the projected population is expected to be above 139 million people by 2020 (Tegegne & Gebre-Wold, 1998). Communities where intervention programmes objectives were met through timely payments of cash transfers have witnessed a significant increase in their cereal production levels. In the year 1999 alone, Ethiopia received 634,774 tons of cereal food aid, which was about 38.2% of domestic cereal production. Cereal food aid shipments decreased to 86,482 metric tons in 2015, which was about 1.8% of domestic cereal production. The benefit of an increase in domestic cereal production is that it automatically reduces Ethiopia's reliance on cereal food aid.

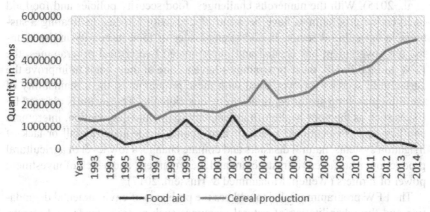

Figure 12.1 Cereal food aid shipment and cereal production in Ethiopia between 1993 to 2015

Source: Authors' computation from FAOSTAT (2019)

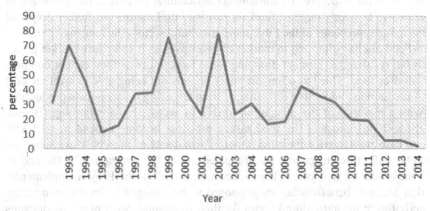

Figure 12.2 Cereal food aid shipment as a percentage of cereal production in Ethiopia

Source: Authors' computation from FAOSTAT (2019)

Sustainability of the Ethiopian food sector

As Endalew and colleagues (2015) posit, more than 5.2 million Ethiopians are food insecure. Ethiopia is naturally endowed with fertile soil, a large workforce, and water resources which have not been fully developed (Awulachew et al.,

2007) or properly managed to improve the agricultural sector significantly. The Ethiopian agricultural sector is prone and vulnerable to weather changes such that when there is insufficient or erratic rainfall for a certain period, there is an increase in the number of people affected by famine in the country (Endalew et al., 2015). With the numerous challenges, food security policies and food aid intervention programmes have been introduced, with a focus on creating a sustainable agricultural sector. In addressing issues of food insecurity in the country, the government has created and implemented two critical programmes, the FFW and the PSNP; these programmes have as one of their goals to improve the agricultural sector. However, even with these programmes, the sustainability of the Ethiopian food sector continues to be questionable. The sector is plagued by issues such as the poor health of farmers, environmental degradation, interethnic conflicts and persecutions, a lack of farm inputs, inadequate rainfall or lack of irrigation systems, natural disasters and climate change, a low level of agricultural production, insufficient infrastructure and local markets, and a lack of investment power of farmers (Welteji, Mohammed & Hussein, 2017).

The FFW programme failed to address the problems of environmental degradation and the rehabilitation of natural resources such as soils, lands, and forests, which are the starting points for a sustainable food system. The adoption of intervention programmes and implementation of food policies has not been effective or problem specific, with untimely transfers of funds to beneficiaries to yield significant results in making household food secured and sustainable (Sabates-Wheeler & Devereux, 2010). Consequently, despite the numerous policies and the continuous massive food aid in the form of cash transfers, households are still food insecure and there are widespread unsustainable agricultural practices. The challenge of excessive or poorly targeted food aid pouring into Ethiopia has further deepened the existing problems in the Ethiopian agricultural sector (Quisumbing, 2003).

Food aid has often led to market disincentives as it can cause market prices to fall resulting in farmers dropping local food prices as they cannot compete with free food. This has often left many farmers not being able to meet the cost of production and feeling discouraged from engaging in large-scale production. With a majority of the population unable to access land for farming (Bogale et al., 2006), most of the food insecure people become dependent on PNSP and cash transfers for their livelihood. Participants of the PNSP are supposed to be self-sufficient or to have graduated out of the programme at the end of a maximum of five years. This is because, based on the Project Implementation Manual, beneficiaries are supposed to be engaged in income-generating activities in the agricultural sector for their livelihood. Most of the participants are unable to be engaged in and dependent on the agricultural sector for their livelihood after benefiting from the cash transfers; this has led to an agricultural sector that is unsustainable.

Conclusion

Ethiopia is challenged with weak and inappropriate government policies, and the country is exposed to severe climatic shocks (droughts) and several interethnic conflicts; these factors have resulted in severe food insecurity and a high dependency on food aid. The number of food insecure people has been rising at a rate of more than 3% per annum over the last quarter century (Berhanu, 2003). Interventions are hindered by the segmented nature and the inconsistent and untimely flow of assistance (cash transfers) to those in need (Nigussa & Mberengwa, 2009). Inclusions and selection criteria for households into food-assistance programmes remain challenged with issues such as corruption, nepotism, amongst many others. These changes necessitate the need for better institutional reorganization and procedures on how interventions can better be done and meeting their objectives.

Additionally, continuous food aid directly affects the labour supply and hence production activities. Households that receive food aid evidently become less productive and shift their time from productive to non-productive leisure activities. This results in a reduction in the labour supply and limits production activities (Abdulai et al., 2004). Another concern is the issue of disguised and open unemployment in rural communities caused by inadequate land for production and lack of alternative employment activities. Land tenure problems remain a concern in the country with no clear-cut policy on land use and ownership to enable those who are willing and able to engage in agricultural activities to do so without any challenges. However, although there is development assistance available in the country, the strengthening of local institutions and policies would go a long way to help the country address its food security challenges.

References

Abdulai, A., Christopher, B., & Peter, H. (2004) *Food Aid for Market Development in Sub-Saharan Africa*. DSGD Discussion Paper No. 5. Washington, DC, International Food Policy Research Institute.

Adal, Y. (2006) Rural women's access to land in Ethiopia, in *Land and the Challenge of Sustainable Development in Ethiopia: Conference Proceedings*, Addis Ababa, African Books Collective, pp. 19–42.

Awulachew, S. B., Yilma, A. D., Loulseged, M., Loiskandl, W., Ayana, M., & Alamirew, T. (2007) *Water Resources and Irrigation Development in Ethiopia*. Working Paper 123. Colombo, Sri Lanka, International Water Management Institute.

Azadi, H., Rudder, F., Vlassenroot, K., Nega, F., & Nyssen, J. (2017) Targeting international food aid programmes: The case of productive safety net programme in Tigray, Ethiopia, *Sustainability*, 9, p. 1716.

Barrett, C. B., & Clay, D. C. (2001) *How Accurate Is Food-for-Work Self-Targeting in the Presence of Imperfect Factor Markets? Evidence from Ethiopia*. Ithaca, NY, Cornell University.

Bassi, M. (1997) Returnees in Moyale district, Southern Ethiopia: New means for an old inter-ethnic game, in Hogg, R. (ed.), *Pastoralists, Ethnicity and the State in Ethiopia*. London, Haan, pp. 23–54.

Berhane, G., Hoddinott, J., Kumar, N., & Margolies, A. (2017) *The Productive Safety Net Programme in Ethiopia Impacts on children's Schooling, Labour and Nutritional Status*, January. Impact Evaluation Report 55. New Delhi, International Initiative for Impact Evaluation.

Berhanu, A. (2003) *The Food Security Role of Agriculture in Ethiopia*. Addis Abeba, The FOA- ROA Research Project at EEA/EEPRI.

Beyene, F., & Muche, M. (2010) Determinants of food security among rural households of Central Ethiopia: An empirical analysis, *Quarterly Journal of International Agriculture*, 49(4), pp. 299–318.

Bishop, C., & Hilhorst, V. (2010) From food aid to food security: The case of the safety net policy in Ethiopia, *Journal of Modern African Studies*, 48, pp. 181–202.

Bogale, A., Taeb, M., & Endo, M. (2006) Land ownership and conflicts over the use of resources: Implication for household vulnerability in Eastern Ethiopia, *Ecological Economics*, 58(1), pp. 134–145.

Coll-Black, S., Gilligan, D. O., Hoddinott, J., Kumar, N., Taffesse, A. S., & Wiseman, W. (2012) *Targeting Food Security Interventions When "Everyone is Poor": The Case of Ethiopia's Productive Safety Net Programme; Ethiopia Strategy Support Program (ESSP II)*. Working Paper 24. Washington, DC and Addis Ababa, Ethiopia, International Food Policy Research Institute (IFPRI).

De Rudder, F. (2013) *Food Insecurity in Rural Tigray (North Ethiopia): Analysis of a Food-Aid Based Response*. Master's Thesis, Ghent, Belgium, Ghent University.

Del Ninno, C., Dorosh, P., & Subbarao, K. (2007) Food aid, domestic policy and food security: Contrasting experiences from South Asia and Sub-Saharan Africa. *Food Policy*, 32, pp. 413–435.

Dercon, S., & Krishnan, P. (2004) Food aid and informal insurance, in Dercon, S. (ed.), *Insurance Against Poverty*, Oxford, Oxford University Press, pp. 305–329.

Devereux, S. (2000) *Food Insecurity in Ethiopia: A Discussion Paper for DFID*. Brighton, IDS.

Devereux, S., Sabates-Wheeler, R., Tefera, M., & Taye, H. (2006) *Ethiopia's Productive Safety Net Programme (PSNP), Trends within targeted households*. Sussex, Institute of Development Studies and Addis Ababa, Indak International Pvt. L. C.

DFID Ethiopia. (2005) *Project Memorandum. Ethiopia: Support to Productive Safety Net Programme*. Addis Ababa, DFID.

Dorosh, P., & Rashid, S. (eds.) (2013) *Food and Agriculture in Ethiopia: Progress and Policy Challenges*. Philadelphia, PA, University of Pennsylvania Press.

Endalew, B., Muche, M., & Tadesse, S. (2015) Assessment of food security situation in Ethiopia: A review, *Asian Journal of Agricultural Research*, 9, pp. 55–68.

Ethiopia-Government. (2013) *Ethiopia: National Policy and Strategy on Disaster Risk Management*. Available at: https://www.preventionweb.net/english/professional/policies/v.php?id=42435

FAO. (2019) *FAOSTAT Database*. Rome, Italy: FAO. [Accessed 11 April 2019].

Gebremedhin, B., & Swinton, S. (2000) Reconciling food-for-work project feasibility with food aid targeting in Tigray, Ethiopia, *Food Policy*, 26, pp. 85–95.

Gebreselassie, S. (2006) *Food Aid and Small-Holder Agriculture in Ethiopia: Options and Scenarios*. A Paper for the Future Agricultures Consortium Workshop, Institute of Development Studies, 20–22 March. Available at: https://assets.publishing.service.gov.uk/media/57a08c45e5274a31e00010ee/SG_paper_1.pdf

Gebru, M., Remans, R., Brouwer, I., Baye, K., Melesse, M. B., Covic, N., Kassaye, T., et al. (2018) *Food Systems for Healthier Diets in Ethiopia: Toward a Research Agenda*. IFPRI Discussion Paper 1720. Washington, DC, IFPRI. Available at: https://cgspace.cgiar.org/handle/10568/98257

Gilligan, D. O., Hoddinott, J., & Taffesse, A. S. (2008) *The Impact of Ethiopia's Productive Safety Net Programme and Its Linkages*. Discussion Paper 00839. Washington, DC, International Food Policy Research Institute (IFPRI).

Gray, C., & Mueller, V. (2012) Drought and population mobility in rural Ethiopia, *World Development*, 40(1), pp. 134–145.

Grepperud, S. (1996) Population pressure and land degradation: The case of Ethiopia, *Journal of Environmental Economics and Management*, 30(1), pp. 18–33.

Jayne, T., Strauss, J., Yamano, T., & Molla, J. (2002) Targeting of food aid in rural Ethiopia: Chronic need or inertia? *Journal of Development Economics*, 68(2), pp. 247–288.

Kaluski, D. N., Ophir, E., & Amede, T. (2002) Food security and nutrition – the Ethiopian case for action, *Public Health Nutrition*, 5(3), pp. 373–381.

Kamenidou, I., Rigas, K., & Priporas, C. V. (2017) Household behavior on food security during an economic crisis, in Mergos, G., & Papanastassiou, M. (eds.), *Food Security and Sustainability*. Berlin, Springer, pp. 243–261.

Kassie, B. T., Rötter, R. P., Hengsdijk, H., Asseng, S., Van Ittersum, M. K., Kahiluoto, H., & Van Keulen, H. (2014) Climate variability and change in the Central Rift Valley of Ethiopia: Challenges for rainfed crop production, *The Journal of Agricultural Science*, 152(1), pp. 58–74.

Kay, S., Taylor, B., & Amdissa, T. (2006). *Targeting Ethiopia's Productive Safety Net Programme (PSNP)*, The IDL Group.

Keulen, H. (2014) Climate variability and change in the Central Rift Valley of Ethiopia: Challenges for rainfed crop production, *The Journal of Agricultural Science*, 152(1), pp. 58–74.

Keller, E. J. (1992) Drought, war, and the politics of famine in Ethiopia and Eritrea, *The Journal of Modern African Studies*, 30(4), pp. 609–624.

Kidane, A. (1989) Demographic consequences of the 1984–1985 Ethiopian famine, *Demography*, 26(3), pp. 515–522.

Levine, S., Crosskey, A., & Abdinoor, M. (2011) *System Failure? Revisiting the Problems of Timely Response to Crises in the Horn of Africa*. Network Paper 71. London, Humanitarian Practice Network.

Lind, J., & Jalleta, T. (2005) *Poverty, Power and Relief Assistance: Meanings and Perceptions of Dependency' in Ethiopia*. HPG Background Paper. London, HPG.

Mariam, M. W. (1986) *Rural Vulnerability to Famine in Ethiopia, 1958–1977*. London, Intermediate Technology Publications.

Markakis, J. (2003) Anatomy of a conflict: Afar & Ise Ethiopia, *Review of African Political Economy*, 30(97), pp. 445–453.

Mitiku, A., Fufa, B., & Tadese, B. (2012) Empirical analysis of the determinants of rural household's food security in Southern Ethiopia: The case of Shashemene district, *Basic Research Journal of Agricultural Science and Review*, 1(6), pp. 132–138.

Mohamed, A. A. (2017) Food security situation in Ethiopia: A review study, *International Journal of Health Economics and Policy*, 2(3), pp. 86–96.

Nega, F., Mathijs, E., Deckers, J., Haile, M., Nyssen, J., & Tollens, E. (2010) Rural poverty dynamics and impact of intervention programs upon chronic and transitory poverty in Northern Ethiopia, *African Development Review*, 22, pp. 92–114.

Nigussa, F., & Mberengwa, I. (2009) Challenges of productive safety net program implementation at local level: The case of Kuyu Woreda, North Shewa zone, Oromia region, Ethiopia, *Journal of Sustainable Development in Africa*, 11(1), pp. 248–267.

Quisumbing, A. S. (2003) *Food Aid and Child Nutrition in Rural Ethiopia*. Discussion Paper No. 158. Food Consumption and Nutrition Division. Washington, DC, International Food Policy Research Institute.

Rahmato, D., Pankhurst, A., & van Uffelen, J. G. (eds.) (2013) *Food Security, Safety Nets and Social Protection in Ethiopia*. Oxford, African Books Collective.

Ramakrishna, G., & Demeke, A. (2002) An empirical analysis of food insecurity in Ethiopia: The case of North Wello, *Africa Development/Afrique et Développement*, pp. 127–143.

Sabates-Wheeler, R., & Devereux, S. (2010) *Cash Transfers and High Food Prices: Explaining Outcomes on Ethiopia's Productive Safety Net Programme*. Working Paper. Sussex, Institute of development studies and center for social protection, University of Sussex.

Segers, K., Dessein, J., Hagberg, S., Develtere, P., Haile, M., & Deckers, J. (2008) Be like bees–the politics of mobilizing farmers for development in Tigray, Ethiopia. *African Affairs*, 108(430), 91–109.

Sharp, K. (1998) *Between Relief and Development: Targeting Food Aid for Disaster Prevention in Ethiopia*. London, Overseas Development Institute.

Sharp, K., Brown, T., & Teshome, A. (2006) *Targeting Ethiopia's productive safety net programme (PSNP)*. London and Bristol, Overseas Development Institute and the IDL Group.

Tache, B., & Oba, G. (2009) Policy-driven inter-ethnic conflicts in Southern Ethiopia, *Review of African Political Economy*, 36(121), pp. 409–426.

Taddese, G. (2001) Land degradation: A challenge to Ethiopia. *Environmental Management*, 27(6), pp. 815–824.

Teferi, L. (2012) The post 1991 inter-ethnic conflicts in Ethiopia: An investigation, *Journal of Law and Conflict Resolution*, 4(4), pp. 62–69.

Tegegne, A., & Gebre-Wold, A. (1998) *Prospects for Peri-Urban Dairy Development in Ethiopia*. Nairobi, International Livestock Research Institute. Available at: https://cgspace.cgiar.org/handle/10568/50997

TGE. (1993) *National Policy for Disaster Prevention and Management*. Addis Ababa, TGE.

Tolossa, D. (2003) Issues of land tenure and food security: The case of three communities of Munessa Wereda, South-Central Ethiopia. *Norsk Geografisk Tidsskrift-Norwegian Journal of Geography*, 57(1), pp. 9–19.

UNHCR (2018) *UNHCR Ethiopia – Nutrition Factsheet*, June. Addis Ababa, UNHCR. Available at: https://reliefweb.int/report/ethiopia/unhcr-ethiopia-nutrition-factsheet-june-2018

UNICEF (2017) *New EU Funding Will Provide Essential Nutrition Treatment for 130,000 Children Under the Age of Five in Ethiopia*. Addis Ababa, UNICEF. Available at: https://unicefethiopia.org/tag/malnutrition/

UNOCHA (2018) *Ethiopia: Oramia – Somali Conflict-Induced Displacement. Situation Report No. 4*. Addis Ababa, UNOCHA. Available at: https://reliefweb.int/sites/reliefweb.int/files/resources/ethiopia_-_oromia_somali_conflict_induced_displacement_june_2018c.pdf

Webb, P., & Von Braun, J. (1994) *Famine and Food Security in Ethiopia: Lessons from Africa*. West Sussex, John Wiley & Sons Ltd.

Webb, P., Von Braun, J., & Yohannes, Y. (1994) Famine in Ethiopia: Policy implications of coping failure at national and household levels, *Food and Nutrition Bulletin*, 15(1), pp. 1–2.

Welteji, D., Mohammed, K., & Hussein, K. (2017) The contribution of productive safety net program for food security of the rural households in the case of Bale Zone, Southeast Ethiopia, *Agriculture & Food Security*, 6(1), p. 53.

World Food Programme. (2011) Food aid flows 2011 report. WFP, United Nations World Food Programme. Available at: https://www.wfp.org/content/food-aid-flows-2011-report

Worku, M., Twumasi Afriyie, S., Wolde, L., Tadesse, B., Demisie, G., Bogale, G., & Prasanna, B. M. (2012) Meeting the challenges of global climate change and food security through innovative maize research, *Proceedings of the National Maize Workshop of Ethiopia*, 3, 18–20 April, Addis Ababa, Ethiopia, CIMMYT.

13 Don't spend more, spend better

Improving the social efficiency of water and sanitation services in Uruguay and South Africa

Adrian Murray and Susan Spronk

This chapter tells two stories, from Uruguay and South Africa, about how grass-roots struggles to improve access to water and sanitation are vital to equitable service delivery. Both stories take place in countries where progressive governments have promised to overcome historical inequalities in service provision and access to water is a constitutional right. Coverage rates are much higher in Uruguay than in South Africa, however. We argue that despite promises to realize the human right to water, grass-roots campaigns help to achieve equity gains in practice. In Uruguay, thanks to the success of a referendum campaign led by workers and community members to make water a human right guaranteed by the constitution, private contracts were reversed, and the public water company has achieved nearly universal coverage not by spending more but by spending better. In South Africa, by contrast, where water is also a constitutional right, local elites have insisted on expensive private sector solutions to recent water shortages and to meet the needs of underserved populations, such as contracting out service extension and delivery and restricting access to the poor.

Keywords

Water and sanitation, right to water, alternatives to privatization, social movements, trade unions

Introduction

Despite efforts by the international community to expand access to water and sanitation, the World Health Organization estimates that 2.1 billion people still lack access to improved water and 4 billion to improved sanitation (UNICEF and WHO 2019). For the past three decades, international financial institutions have insisted that since the public sector has failed to deliver services, the answer to the problem is privatization (World Bank 2018; Bakker 2010). This chapter argues that this logic is misguided. Public and private water and sanitation providers face similar challenges: high upfront costs are required to deliver services to the poor. In order to make this argument, we tell two stories, one offensive and one defensive, that present alternatives to privatization but also aim to improve public service delivery.

What both case studies demonstrate is the importance of pushing the boundaries of the concept of "efficiency" to focus on more social criteria rather than just narrow financial ones when evaluating the performance of water and sanitation utilities. For public services such as water and sanitation, the benefits to the economy and society are largely accrued outside of the provision of services – that is, in better public health outcomes, which in turn have a variety of positive implications for well-being and productivity. Delivering water and sanitation services to poor communities, which have traditionally been excluded from public and private systems, does not require that governments just spend more, it also requires that they spend better. Most importantly, they must invest seriously in achieving service equity.

The concept of social efficiency helps us to rethink the way that we fund and deliver public services to promote social citizenship, based upon a robust concept of human rights. The idea of social efficiency challenges the notion that the private sector is inherently better than the public sector, pointing out that it is difficult to compare the "efficiency" of public and private providers. The primary goal of a private corporation is simple: to make profit; it is accountable only to its shareholders. By contrast, the goals of a public company are much more complex. The government is accountable to its citizens. But a political difficulty remains. If the poor do not vote, why would a politician spend a lot of time and money trying to bring them services?

At base then, we argue that the struggle for water and sanitation is connected to struggles for an expanded notion of citizenship that recognizes that we are all created equal and deserve equality of opportunity. The right to water entails that the state has a responsibility to provide these services when the private sector is not capable of or interested in doing so at an affordable price. The pricing of water and sanitation services must be based upon ability, not willingness, to pay. Wealthy people should not have the "right" to fill their swimming pools and wash their cars just because they can afford to do so. Poor people should not be cut off because they cannot afford to pay their water and sewer bills.

The chapter is organized as follows. The first section of the chapter presents the case study from Uruguay, telling the story of how a successful grass-roots campaign to reverse a private water contract and make water a human right has had long-lasting effects on improving the performance of the publicly owned and operated utility by involving unionized workers in a reform programme. The second section of the chapter presents the case study from South Africa, telling the story of how grass-roots social movements had to fight to defend access to water in the context of the Cape Town water crisis, despite the fact that access to water is constitutionally guaranteed in the country.

This chapter is based upon qualitative, primary research with key informants sponsored by the UN-Global Water Operators Partnership Alliance and conducted by researchers affiliated with the Municipal Services Project in July 2015, as well as ethnographic fieldwork conducted from 2015 to 2018 with the Housing Assembly in Cape Town around access to water more generally and the water crisis in particular.

Social efficiency in Uruguay

Uruguay is one of the few countries in Latin America where most citizens turn on the tap, fill a glass of water, and drink it without thinking twice. The Obras Sanitarias del Estado (Sanitation Works of the State, OSE), has been providing high-quality, affordable, nearly universal water services to the population of Uruguay since 1952. Due to high levels of public investment over six decades, OSE has achieved amongst the highest coverage rates for water and sanitation in the region. About 96% of Uruguayans have access to an improved water source, and 78% have access to improved sanitation. The public health benefits are apparent. Uruguayans consider it a point of national pride that it was the only country on in South America not to be affected by the cholera epidemic in the late 1990s.

OSE is a model public company that promotes the values of social efficiency. The public company has many programmes to increase access and improve equity. These include a programme to provide subsidies to households in informal settlements that connect to public water and sanitation. It also employs the residents themselves, creating much-needed jobs. It also has a rural extension programme to connect schools and households in agricultural areas. Why has OSE been so successful while other public utilities have failed?

One of the reasons that OSE has been so high performing is that OSE workers and the citizens who use the services they provide believe that access to water is part of their citizenship rights. When you talk to the workers at the public utility, one of the first things you might notice is how proud they are to work there and to serve their community. This is a story of how the community of workers and citizens got together to fight to return the utility to public control, and how this fight has helped push for better management, which saved enough money to create new programmes to reach marginalized communities. The social movement campaigns to exert public control over the utility have helped the utility not spend more but spend better.

History of OSE: a strong public ethos

OSE was established by the central government as a publicly owned and operated water and sanitation company that operates at arm's length from government. OSE is widely considered by social movement activists within the region to be a model public company. Within OSE, the culture of public ethos is so strong that there is even a term for it. Workers feel a strong emotional connection to the public water company and identify themselves as "*camiseteros*" ("shirt-wearers"); in other words, they "wear the shirt" of the public water company in the same way that they wear a shirt to express loyalty to their football team. And given the seriousness with which Uruguayans take their football, such a statement should not be taken lightly. As Carolina Moll, a human resources manager describes (author interview), workers at OSE feel that there is a strong relationship between citizenship and state:

> If you go to a small town in Uruguay, the OSE worker is the one who raises the flag in the morning, answers questions of the citizens, hands out the water

bills. And in the afternoon, the OSE worker lowers the flag, folds it up and puts it away. It is a form of love of country.

The close ties amongst workers also explain the high level of "public ethos" within OSE. Workers and managers describe OSE as a "big family." Many trace their OSE lineage back to their grandparents and great-grandparents. Adriana Marquisio (author interview), one of the key spokespersons for the Red Vida (Latin America's largest anti-privatization network) and former president of the trade union the Federation of Public Servants of Sanitation Works of the State (Federación de Funcionarios de las Obras Sanitarias del Estado, FFOSE), describes with pride how she is the "third generation" of OSE workers in her family. Her grandfather also worked at OSE. She entered the company as administrative support staff over 30 years ago and eventually worked her way up at one point to be Communications Advisor in OSE, one of the highest paid positions in the company. Carolina Moll (author interview), who has more than 30 years of service with OSE, tells a similar story:

> There are a large number of civil servants with OSE, especially the older ones, who identify strongly with the company. . . . My father started working for OSE in 1960. . . . When I went to find a job, I was an 18-year-old university student. It was difficult to find work because we were in the middle of the dictatorship. In 1980, I applied for a job in a competition for the family members of civil servants and I got my first job. . . . OSE taught me everything. Sometimes we complain because our salaries are low, but a public job has real advantages – it allowed me to continue studying and build my career.

This depiction of OSE as a close-knit community based on strong familial ties defies conventional beliefs about the problems of nepotism and corruption that plague public utilities. As Doaiz Uriarte (author interview) explains, one of the reasons that corruption is so low in OSE is that "everyone is the relative of someone." Problems related to nepotism are mitigated by the rule that no two family members can work within the same department. He emphasizes as well that low rates of corruption are not unique to OSE but relate to Uruguay's tradition of wage equity in the public sector: the president of the Republic earns a salary of US$10,000 per month; a senator, US$6,000; and a minister, US$5,000, while the average worker in OSE earns US$2,000. Uruguay has one of the highest wage equality rates in Latin America.

The role of workers and communities in public service reform

One of the other reasons that the public ethos in OSE is so high is the protagonist role that the trade union has played in the operations, management, and defence of the public company. Indeed, it is impossible to tell the story of OSE's return to public control without telling the story of the water workers' union, the Federación de Funcionarios de las Obras Sanitarias del Estado (Federation of Public Servants of Sanitation Works of the State, FFOSE).

In the 1990s, the World Bank recommended to the Uruguayan government to privatize part of the water services as a means to "modernize" the sector by introducing competition. Following this advice, in 1992 the Uruguayan government granted the first concession for water and sanitation services in the wealthy, eastern zone of Maldonado to the company Aguas de la Costa. In 1998, FFOSE began to manifest its opposition to privatization as rumours began that the government planned to increase private participation in the sector. Despite growing public opposition, OSE granted a concession for the entire department of Maldonado to URAGUA, a subsidiary of Spanish water company Aguas de Bilbao in 2000.

In October of that year, various organizations founded the Comisión en Defensa del Agua y Saneamiento de la Costa de Oro y Pando (CDASCOP) – a precursor to the Comisión Nacional en Defensa del Agua y la Vida (National Commission in Defense of Water and Life, CNDAV), which would become the main coalition of social movement organizations that opposed privatization of water and sanitation. Residents of de la Costa held a series of demonstrations in front of OSE's offices in the cities of de la Costa and Canelones. OSE workers held partial work stoppages in solidarity with these protests. Along the way, dozens of organizations joined the campaign including the original CDASCOP and the water workers' union (FFOSE), as well as the environmentalist groups Liga de Fomento de Manantiales and the Red Ecológica Social-Amigos de la Tierra (Friends of the Earth). Each of these organizations brought a distinct but complementary vision to the campaign: neighbourhood organizations stressed the importance of access to affordable services, ecologists brought an environmentalist perspective, and the workers shared their concerns about the impact of privatization on their working conditions and service delivery.

FFOSE is widely considered the "backbone" of the social movement alliance that made water a human right, providing ideological direction, and human and financial resources. In 2000, FFOSE published a series of pamphlets that laid out the ideological basis for the CNDAV campaign to push for a national referendum to promote the right to public participation and the return of the service to public control based upon the notion that water as a "fundamental human right." The referendum aimed to modify the constitution to declare that "all superficial and subterranean water resources in all of its forms and uses are not alienable to multinational capital."

The campaign to collect signatures to call for the referendum was officially launched on October 18, 2002. The objective was to collect 250,000 signatures. Within a year, 282,000 signatures were collected, and the campaign shifted gears to mobilize the "yes" vote. The coalition organized months of national and international events to draw attention to the campaign. Activists from all over the continent came to Uruguay to speak in favour of the campaign, along with local celebrities. On October 31, 2004 – the same date as the electoral victory that brought the new left-of-centre government to power – 64.7% of the population voted to make water a human right, a victory which set a fundamental precedent in the defence of water and for direct democracy.

When the new left-of-centre government, Frente Amplio, came to power in 2005, it also signified a change in political orientation in the public water company. The new board of directors faced the challenge of making the "human right to water" a concrete reality. When the utility was returned to public control in 2005, it was in bad financial shape. As Carolina Moll and Doaiz Uriarte (author interviews) explained, OSE's deficit skyrocketed following the regional economic crisis triggered by the Argentine collapse in 2002. In the context of a regional economic crisis precipitated by the collapse of the Argentine economy, some workers were reduced to partaking in petty corruption to augment their income, such as charging the company US$600 for "travel expenses" when they did not even leave Montevideo. Another and arguably more important factor was the concession in Maldonado, which bled the utility of a major source of tariff revenue. Due to the poor financial state of the utility, the Inter-American Development Bank refused to grant OSE a loan in the early 2000s (Spronk, Crespo and Olivera 2014). They had to do something.

Buoyed by the success of the constitutional campaign and electoral victory, the new board of directors – many of whom were involved in the campaign as members of FFOSE and appointed by the Frente Amplio government – executed an ambitious plan to rid the company of corruption and eliminate wasteful practices. New internal accounting mechanisms were established to place stricter controls on spending, particularly for employee travel, and to restrict the use of company vehicles to official business. To weed out workers engaged in abusive practices, OSE conducted 50 internal investigations, which resulted in 14 prosecutions of former directors (Spronk, Crespo and Olivera 2014). Crucially, this painful restructuring process involved the active support of FFOSE, which ensured that workers received due process but would not and did not defend workers found guilty of conducting illicit activities that harmed the company. These reforms increased accountability, and the company, which was running an operating deficit of US$3–4 million per year in 2004, turned its financial situation around. Today, OSE runs operating surpluses. Importantly, efficiency has been achieved without sacrificing social goals, making OSE a model public water company of social efficiency.

A model public water company: extending services to the poor

Over the past six decades, OSE has built relationships of solidarity not only amongst the workers of the company but also the community. The fact that OSE is a *national* water company has allowed Uruguay to achieve economies of scale and enact a tariff policy that redistributes wealth from high-income to poor users and transfers resources from urban to rural areas. OSE's expansion plan demonstrates that it takes seriously its responsibility to achieve its mission: to contribute to public health and its socio-economic development by providing potable water and sanitation to every household in country.

OSE uses a rising block tariff system of cross-subsidization, which is the most progressive form of charging users for water. The tariff structure distinguishes between 26 categories. Commercial and industrial customers are charged more per litre than domestic customers, and tariffs vary depending on the zone, diameter of water pipe, or type (or absence) of water meter. The charge for water rises substantially by 5m³ blocks beyond consumption of 15m³. Impressively, the utility is auto-financing: the rising block tariff system allows the company to cover its maintenance and operation costs. The tariffs from Montevideo make up 67% of the company's revenue, which is used to subsidize rural users. In past years, hikes to the water tariff have been below the rate of inflation, which means that the price of water has effectively dropped (Spronk, Crespo and Olivera 2014).

Households in the informal settlements consume higher volumes of water than in other parts of the city, which raises a potential concern about the equity impact of the rising block tariff. According to one report, on average, households in the settlements in Montevideo consume 25m³ of water per month, almost double the average consumption of a household in Montevideo, which is only 13m³ per month. As Gabriel Apolo (author interview), the manager of the Program to Reduce Unaccounted for Water, explains, higher consumption rates are due to more populous households and leaky infrastructures, the majority of which are self-constructed. In the *asentamientos* it is not an uncommon sight to see water running down the streets due to leaks in the pipes.

This potential negative impact of the rising block tariff is mitigated by an effective system of public subsidies. In 2005, the Frente Amplio government created the Social Development Ministry (Ministerio de Desarrollo Social, MIDES), which was charged with the task of implementing the Frente Amplio's flagship programme, the Social Emergency National Assistance Plan (PANES), which targets the 10% of households living in extreme poverty (Lanzaro, 2011). The average household pays about US$20 for water per month, while households registered with the MIDES pay a "social tariff" of US$2–3 per month regardless of their consumption. In 2009, there were about 25,000 households (about 3% of OSE's total customers) that were registered with the plan (MIDES, 2009).

The quantity and quality of water services provided to households with OSE water connections are also excellent. Water is available 24 hours per day, 7 days a week. While water demand rises in the hot summer months, the utility does not suffer from water shortages. Since the water taking at the Santa Lucia River – which provides Montevideo with 100% of its drinking water – is a secure resource that drains from one of the world's largest aquifers, scarcity is not a pressing public issue. Even more importantly, the water and sanitation infrastructure in Uruguay has been built over decades. When there are occasional problems with water quality, OSE puts a drinking water advisory in effect. For example, when a rupture in a water main threatened the safety of Montevideo's water supply for a brief period in 2007, OSE distributed, to households for free, bags of drinking water that were produced by the company's own water packaging plant. It is a company that is deeply committed to the welfare of the community.

In sum, the OSE story demonstrates that public utilities can and do perform, especially when managers have close ties with the community. Too often, reform projects in public utilities depend on cheapening the wage bill by firing full-time workers and hiring them back on contract. The story of OSE demonstrates that restructuring can be done differently, and with buy-in from the trade union. It is possible to maintain fair working conditions, which helps inspire loyalty amongst the workforce. The OSE story also demonstrates that it's possible to extend the network to marginalized communities when you price water in a fair way, based upon ability, not willingness, to pay.

The Cape Town water crisis and the politics of social movement defence

South Africa has made enormous strides in extending basic services to its population in the post-apartheid period. According to official statistics, over 80% of households have access to safe drinking water and electricity, and more than two-thirds have improved sanitation (StatsSA, 2016). Despite the remarkable improvements suggested by these numbers, access to services is undermined in practice by the commodification of services, the corporatization of public utilities, the proliferation of cost-recovery mechanisms, and poor service quality in the context of fiscal austerity. In other words, access to services is increasingly mediated by one's ability to pay. All of these factors limit the ability of South Africans, the poor working-class especially, to benefit from being connected to this greatly expanded infrastructure. The roots of these barriers and a recent deterioration in access to many services in the country can be located in an adherence to narrow principles of financial efficiency in the post-apartheid period.

Following the fall of apartheid in 1994, expectations for a new South Africa and the possibility of the "better life for all" promised by the African National Congress (ANC) were high. The 1995 Reconstruction and Development Plan and the 1996 Constitution – which guaranteed access to water and was widely hailed as the most progressive in the world at the time – appeared to provide a progressive legislative agenda and policy programme for eradicating the inequalities and addressing the injustices of apartheid. This transformative vision was diluted when the ANC government initiated a plan known as Growth Employment and Redistribution (GEAR) in 1996. Described as a self-imposed structural adjustment, GEAR and subsequent legislation restructured the economy along market lines, which seriously compromised the promise of universal service delivery.

Bearing the mark of the World Bank's neoliberal urban orthodoxy, the legislative agenda that began with GEAR downloaded the responsibility for service provision onto local governments while simultaneously imposing fiscal austerity. This led to a crisis of service delivery as local governments sought to corporatize and privatize services in an attempt to achieve financial efficiency. Millions of people had their water and electricity disconnected in the late 1990s (McDonald, 2002). Large-scale "service delivery" protests erupted by the early 2000s.

The state responded by increasing expenditure but tightening their grip on the poor by restricting their access to entitlements. Although free basic services (FBS) have been provided to "indigent" households since 2001, side by side with a steady investment in the coercive arm of the state, service provision levels remain well below international standards for basic subsistence in many parts of the country. Access is particularly poor in working-class areas and burgeoning informal settlements in cities and towns and for the shrinking rural population (Marais, 2012). Stemming from this lack of access, among other issues, South Africa continues to have some of the highest levels of protest in the world.

The Housing Assembly

The Housing Assembly is a coalition of community organizations primarily from Cape Town and the surrounding area that has been struggling for better access to housing and related public services for nearly a decade. The Housing Assembly seeks to challenge, both materially and ideologically, the conditions of everyday life for poor, working-class communities and to engage in the process of building alternatives to the privatization of public service delivery. They also fight other forms of commodification which increasingly characterize daily life in South Africa. For many years the Housing Assembly has challenged the commodification of water, specifically fighting against the installation of water management devices (WMDs), which are a key component of the local government strategy to limit both water use and as a means to try and address the ongoing fiscal and service delivery crisis in not only Cape Town but in municipalities across the country. The Housing Assembly's efforts became all the more urgent in 2017 when a prolonged drought threatened the city's water supply.

The water crisis

Water levels in the reservoirs and dams supplying Cape Town reached critically low levels in 2017, and late in the year the city warned that it could run out of water, or reach "Day Zero," in early 2018. Water consumption was dramatically reduced through several measures: household and eventually commercial conservation, pressure reduction, the installation of household flow regulators – aka, water management devices (WMDs) – and punitive tariffs. From the perspective of state managers, the measures "worked": consumption fell from 900 million litres per day (MLD) in February of 2017 to 500 MLD in February of 2018 (CoCT, 2018a).

As a part of the FBS initiative, indigent households receive up to $10.5m^3$ per month of water for free. At the height of the crisis in early 2018, all households were instructed to reduce consumption to below $6m^3$ litres per month (or 50 litres per person per day based on a four-person household). If usage exceeded $10.5m^3$ litres per month (87.5 litres per person per day), households faced the installation of a WMD to manage and cut-off the flow of water. On top of this, a tariff hike of 26.9% (along with a similar increase for sanitation) was introduced in January 2018, which ramped up after $6m^3$ litres per month (CoCT, 2018a).

Despite the fact that several state reports published as far back as the early 2000s had warned of a looming water crisis and urged various levels of government to take action to invest in infrastructure, reduce non-tariff water – leaks, in other words – improve management, and secure new sources of supply, the city insisted that it was "caught off guard" by the crisis. Instead, in a press release in early January 2018 the city shifted the blame firmly onto the shoulders of residents, which it characterized as careless and unwilling to save water, adamant that it would proceed to force wasteful Capetonians to save water through the aforementioned measures (CoCT, 2018b). But as one city councillor and former executive deputy mayor pointed out in a speech to city council at the time, the city's claim that the crisis came as surprise was "complete nonsense."

The city's narrative was particularly infuriating for working-class Capetonians, who use disproportionately less water than more wealthy residents. For example, informal settlements, which usually receive water from free-flowing communal standpipes, account for only 4% of Cape Town's water consumption despite making up about 15% of the city's population (CoCT, 2018a). Unlike in Uruguay, where household size and ability to pay are considered in water provisions to poor communities, poor families in Cape Town who live in formal housing almost universally find that the 10.5m^3 per month of free water is insufficient due to the large size of working-class households, in which as many as 15 to 30 people end up sharing the free water allocation.

Meanwhile restrictions on "pool refills" were not put in place until dams supplying water to the city were at approximately 15% capacity. Despite these vastly unequal conditions of access and consumption of water, there were numerous reports of wealthy and middle-class residents blaming the poor for the crisis Housing Assembly 2018.

The city's thoroughly neoliberal response to the crisis echoed this sentiment in its disproportionate focus and impact on working-class communities, and this has only served to deepen inequality in Cape Town. This focus is evident in the disproportionate installation of WMDs in working-class communities – where more than two-thirds of total installations took place in the first quarter of 2018 – and restrictions to water access in informal settlements. These measures led to severe water shortages, widespread health problems, and even deaths when there was no water to extinguish the widespread shack fires that rage across the city's informal settlements each summer. These negative ramifications of the city's response are also clear in the disproportionate impact of increases in water tariffs on working-class household budgets, which are already stretched to the breaking point, and the disproportionate impact of the crisis on the quality of working-class life.

As many working-class women noted in the course of the crisis, despite the fact that in situations of inadequate housing in which the majority of the working-class lives more not less water is needed to meet daily needs, getting by on 50 litres a day is nothing new. The city's response to the crisis just made this reality worse, and more expensive. This market mediation of access to water, which the response to the water crisis has greatly intensified, has a disproportionate impact on women, who do the majority of gendered household and community labour. The reality is that working-class women bore the brunt of the crisis, as they were forced to manage escalating restrictions on water and household finances in a situation already characterized by inadequate access to water; water that is cut off every day by a WMD after 360 litres has been dispensed; water that must be fetched from a communal standpipe several hundred metres away; water that must be negotiated for with the landlord; water that must be recycled over and over and used for many tasks; water that must be conserved, so children are told to not wash their hands and not flush the toilet. The public health implications, among others, are glaring in the way water restrictions impact everyday life. All the while, little has changed for wealthier residents. As one middle-class resident of Vredehoek, an upscale area close to the city centre, told me, "I shower at the end of the day now, rather than at the beginning too."

The Cape Town Water Crisis Coalition and the Housing Assembly

But despite the thin reality of the material conditions of everyday life in Cape Town, the city's working-class residents have not accepted the city's scolding silently nor have they taken the response to the crisis laying down. In early January 2018, when the City of Cape Town (finally) welcomed labour and community organizations, ratepayer and tenant groups, and businesses and other civil society actors into the council chambers to draw to a close a lacklustre public participation process around amendments proposed to the water bylaw – necessary to implement the series of aforementioned measures to deal with the crisis – they were given an earful. From the impassioned conversations that followed, the Cape Town Water Crisis Coalition (henceforth, the Coalition), was formed and launched the following week by a number of community, labour, and faith-based organizations from across Cape Town, the Housing Assembly among them. A series of committee structures – organizing, media, research, and so on – were quickly formed. In the following weeks, through local and citywide direct action and engagement with the city, national government, and businesses, a variety of small victories were won. These victories included the physical blocking of WMD installations, the opening of public and private springs for public water consumption, and the dispelling of myths around the cause of the crisis and who was to blame.

While the bulk of its member organizations are from the poor, working-class townships of the Cape Flats, the voices of middle-class members initially had a disproportionate impact on the Coalition's strategy and tactics. This led to an early focus on measures which failed to address the working-class experience of the crisis and the historical roots of inequitable service delivery. For example, the springs that were opened were all in middle and upper-class areas, out of the reach of the majority of Cape Town's working-class residents. Technical and legal responses to WMD installations and rates increases, which the Coalition was ill-equipped to pursue and proved unfruitful beyond the very short term, also distracted from the urgent need to organize communities to take direct action to block the installation of WMDs in the first place.

Housing Assembly organizers and their working-class comrades challenged this trajectory and refocused the Coalition on organizing in working-class communities and developing a broader critique of water commodification based on the working-class experience of access to water and/or lack thereof. They also demanded that the city stop the installation of WMDs, especially in working-class areas, and address the crisis through investment in public infrastructure, sustainable management, and equitable distribution of water resources, and institute more progressive tariff structures (unlike Montevideo's progressive 26 block tariff structure and generous cross subsidization, Cape Town has only four residential blocks and a single relatively low tariff for businesses) and take steps to bring more water resources back under public control.

In the words of a Housing Assembly (2018) statement in early 2018:

> As water campaigners from below we were fighting . . . to take the struggle to the masses, people are suffering with the water meters, high bills, etc. We continued to raise the struggle and resistance against the meters. This led to the Coalition attending and arranging meetings closer to working class communities, as the meetings in the Central Business District have excluded many.
>
> These meters are not only faulty but installed by workers who know nothing about plumbing [author's note: a result of the subcontracting process and the use of the cheapest labour]. Often in our communities, after the meters have been installed, families are without water for days and in some cases for weeks.
>
> The working class are saying No to Water Meters! We do not want our water to be privatized, we know the metering system is privatization. Communities across the city have been getting organized, water crisis committees have been formed in various areas across the city [author's note: community-run advice offices have also been created], there have been marches to municipal offices, pickets and other activities, communities have also been resisting and chasing away the installers of the water management devices.

As one Housing Assembly (2018) organizer described, reflecting on a meeting in February 2018, a very simple strategy was utilized in organizing meetings: "The meeting put together a leaflet [which] highlighted the issues people are faced with. A questionnaire was designed to collect cases to expose how defective the [water management] devices are and how the limited amount of water impacts their daily lives." Rather than getting bogged down in the technical and legal confines of neoliberal water governance and service delivery, Housing Assembly organizers engaged with the daily experience of working-class water access in an effort facilitate critical reflection, community organizing, and direct action to challenge the city's agenda. Connecting the dots between the reality of everyday life and the broader context of water commodification and privatization were key to this organizing effort.

This organizing approach is one of the reasons why the Housing Assembly has been successful in drawing together a loose working-class coalition within but increasingly beyond the Water Crisis Coalition. This approach works against commodification in both the short and long term by articulating a position in which the push to commodify services like water is framed as a part of the state's and capital's neoliberal agenda to keep spending low and provide new avenues for productive investment. Indeed, during the crisis it came to light that members of the then-mayor's family had received contracts for WMD installations in an outsourcing process that lacked transparency.

An important part of the way the Housing Assembly counters this narrative of scarcity and austerity is to lay bare the inequitable ownership, management, and distribution of water, and demand not only more spending on access to services in working-class communities, but better spending. Particularly poignant is the Housing Assembly's call for prioritizing investments in improving public infrastructure and service delivery rather than relying on expensive, often capital intensive, private sector solutions and rampant outsourcing, which has proven to be both socially and financially inefficient, lining the pockets of the well-connected few at the expense of the many.

Together these arguments emphasize the necessity of the patient work of organizing; organizing of the sort practiced by the Housing Assembly, which prioritizes the experience of everyday working-class life as a starting point from which to build broad working-class social movements which are key to ensuring the practical realization of constitutional rights like the right to water.

Conclusion

This chapter tells two contrasting stories of local actors that were motivated to make the right to water a concrete reality.

In the case of South Africa, despite the fact that the post-apartheid constitution is one of the most progressive in the world, fiscal constraints have impeded the realization of the right to water guaranteed in the constitution. Social movement organizing, like that carried out by the Housing Assembly and the Water Crisis Coalition, has physically defended the right of access to water in practice and has blocked the city's attempts to further commodify and restrict access to water as a way to resolve the supposed crisis. The crisis itself, we argue, was not caused by water scarcity but rather the construction of scarcity due to inequitable ownership, management, and distribution of water resources. In other words, there are many affordable solutions to this crisis, including investing in infrastructure and publicly owned and operated service delivery rather than relying on expensive technologies and rampant outsourcing, which elevate costs and line the pockets of the local elites.

In Uruguay, by contrast, the sustained and successful leadership of organized workers and community members led to a turnaround in the public water and sanitation company, one that helped it save money and realize the substantive promise of the human right to water by investing in projects to bring water to the poor. Importantly, the campaign for constitutional reform was a bottom-up effort and helped create buy-in from workers who were keen to reform operations in order to keep the company public. Their collective efforts helped to clean up a corrupt public water company and forge a new path to realize social efficiency.

These stories demonstrate how the right to water must not be constrained by marketized notions of rights and efficiency that focus on narrow financial criteria and which inevitably result in inequitable outcomes since the wealthy can afford to waste water by filling swimming pools just because they can afford to pay for it. These stories also demonstrate that having progressive governments and constitutional guarantees of rights – as is the case in both Uruguay and South Africa – are not enough either. In order to make governments and service providers accountable, we need sustained, grass-roots, collective action by communities and workers to make our vision of social efficiency a reality.

References

Bakker, K. J. (2010). *Privatizing water: Governance Failure and the World's Urban Water Crisis*. Ithaca, NY: Cornell University Press.

City of Cape Town (CoCT). (2018a) *Water Outlook 2018 Report*. Cape Town, DWS.

City of Cape Town (CoCT). (2018b) *Day Zero Now Likely to Happen – New Emergency Measures*. Cape Town, CoCT Media Office.

Housing Assembly. (2018) *Water Crisis Coalition Report*, March. Cape Town, Housing Assembly.

Lanzaro, J. (2011). Uruguay: A Social Democratic Government in Latin America, in S. Levitsky and K. M. Roberts (Eds.), *The Resurgence of the Latin American Left*. Baltimore, MD: John Hopkins University Press.

Marais, H. (2012). *South African Pushed to the Limit: The Political Economy of Change*. London, Zed Books.

McDonald, D.A. (2002). The bell tolls for thee: Cost recovery, cut offs, and the affordability of municipal services in South Africa, in McDonald, D. A., & Pape, J., (eds.), *Cost Recovery and the Crisis of Service Delivery in South Africa*. London, Zed Books, pp. 161–179.

MIDES (2009). Lo que toda uruguaya y uruguayo debe saber sobre el MIDES. Montevideo, Uruguay, Ministerio de Desarrollo Social.

Spronk, S., Crespo C., Olivera M. (2014). Modernization and the boundaries of public water in Uruguay. In D.A. McDonald (Ed.), *Rethinking Corporatization and Public Services in the Global South* (pp. 107–35). London: Zed Books.

Statistics South Africa (StatsSA). (2016) *General Household Survey*. Pretoria, RSA.

UNICEF and WHO (2019). 'Progress on household drinking water, sanitation and hygiene, 2000–2017.' New York: United Nations Children's Fund (UNICEF) and World Health Organization. Accessed August 16, 2019 at https://data.unicef.org/wp-content/uploads/2019/06/JMP-2019-FINAL-high-res_compressed.pdf

World Bank (2018). 'From Billions to Trillions: MDB Contributions to Financing for Development.' Washington, DC: The World Bank. Accessed August 16, 2019 at http://pubdocs.worldbank.org/en/69291436554303071/dfi-idea-action-booklet.pdf

Part VI
Global South in the Global North

14 Indigenous peoples in Canada – the case of Global South in the Global North

Part VII

Social capital

The highest form of capital

15 Conclusion
Can a single spark create a prairie fire?

Fayyaz Baqir

The concluding chapter will cast a glance at the thought barriers in the way of local development and practical solutions offered by the better spending principles described in the case studies that deal with these barriers. In view of local successes achieved, the chapter will explore policy implications of better spending strategies for development effectiveness.

Keywords

Social capital, sustainable development, better spending, resilient communities, trust deficit

Thought framework constraining local development

Local development is not a reproduction of global development on a small scale. Global development should not be seen as the elimination of the diversity and richness of the local choices, promotion of uniformity of goals, and denial of the resourcefulness of the local communities. Local development is a re-appropriation of the intellectual, social, and economic space encroached upon by the global systems in the name of their universal validity. Supporting local action does not mean the appropriation of the local in the name of improvement. Local Action is a recreation of the stifled expression of the unique and vibrating community (Groucher, 2004). This expression is closely connected with the acceptance of diversity, plurality, and respect for small enterprises. Contemporary thought patterns shake the foundations for the development of the local in invisible and imperceptible ways. These patterns belittle the creativity, resourcefulness, and power of local action in enriching, informing, and achieving the global goals. Better spending practices have provided us an opportunity to enhance our understanding of the interdependence between the local and the global.

Locally inspired, financed, and managed action has the potential of transforming local communities and generating a ripple effect. In contrast, globally conceived and promoted action may not continue after the end of the project period. Because of the paucity of resources, low-income local communities use their resources very judiciously and find solutions for their problems within their means. They blaze the trail for others. They send the message that "anyone can make the difference." These projects are guided by "frugal sciences" building on a local body of knowledge, and they represent the "generative themes" in each community (Radjou et al., 2012; Freire, 1970). Their key principle is not asking for more dollars but stretching the dollar. Better spending practices don't underrate the importance of "thinking globally and acting locally" but redefine the relationship between the local and global as interactive, dynamic, and complementary (Stephen, 2004). These examples also indicate that better spending and appropriate spending are self-sustaining forms of more spending.

It is important to note here that surveying, costing, designing, budgeting, financing, delivering, and maintaining services through locally implemented programmes can decrease the service delivery cost significantly compared to the cost of delivery through donor, state, or market sector arrangements in the formal sector. Field experience tells us that if spending is done in line with local conditions and resource endowments, it can increase the spending effect by a few hundred percent without allocating any additional resources. Some reasons for this high impact are due to the fact that formal sector agencies in most cases do not have maps and inventories of local physical assets; their service designs are costly because they have been developed in line with the social conditions of higher-income classes; their tendering and contracting process increases the cost of delivery; and their projects are executed through international or national consultants whose services are much more expensive compared to local-level professionals who can deliver services much below the "global" norms with a little bit of guidance.

Low human and social development indicators at the local level are also attributed to weak political will, corruption, and the clash of interests between communities and the power structure. This interpretation does not consider the complete picture. It neglects the fact that there are sizable unused and wasted resources allocated by the state but not accessed by the low-income communities. In most of the cases, waste of resources is caused by a lack of trust between the communities and formal sector experts, and their inability to cooperate to find a solution based on shared understanding. Financial support for finding improved and sustainable solutions does make the differences, but it depends to a large extent on trust building. It is true that local elites give low priority to improving the of quality of life of local communities, but cooperation between the communities and elites is not a zero-sum game (Wright, 2001). Both sides win if available resources are used effectively. Effective spending in such a situation is an appropriate form of better spending.

The zero-sum perception contributes to social mistrust and cannot be eliminated with more money but better understanding and better spending. Elites need to know that there is enormous improvement possible within the given resources if the solutions are developed in line with the local condition, and if social sector expenditure is seen not as subsidizing the poor but investment in their capacity to share the burden for improving their quality of life. Local action also needs to be seen as a way of mobilizing enormous resources through the engagement of local partners and can diminish the need to seek more funds through external economic assistance. Local enterprises also have a smaller carbon footprint, and the success of local development initiatives can contribute to decarbonization of the economy in a big way. Local action paves the way for reducing income inequalities and mitigating climate change by helping communities re-appropriate the economic space from the state and market sectors.

Lessons from the field; the science of counter-intuitive learning

Case studies presented in this volume provide us many counter-intuitive lessons. Daniel Kahneman has made a case for counter-intuitive learning by looking for numbers instead of appearances in search of understanding (Kahneman, 2011). He said that intuitive understanding is convenient and fast because it looks at the similarities between two different phenomena and ignores the differences. It is very helpful in carrying out our daily chores. But it also leads to stereotyping and may obscure critical differences between two events. An important method to overcome the limitations of intuitive judgments is to look for numbers. Numbers help us find counter-intuitive and clear answers to important puzzles. This counter-intuitive method is also very useful in understanding the relationship between global goals and local action. When we compare global and local solutions for globally designed goals, we gain deeper insights by looking at the numbers. One important insight is that expensive global solutions can be replaced with frugal local solutions, and locally implemented inexpensive solutions can appear very expensive to global development experts. Global project budgets are considered low-cost if they have low overhead costs; that is made possible by dealing with a few large-scale projects, and the overhead cost becomes prohibitive if the budget has to cover a large number of small locally implemented projects. On the contrary, high-cost international implementation reduces overhead expenses but substantially increases the cost of service delivery to local communities compared to their capacity to pay back.

When we take a counter-intuitive glance, we find out that the problem of better performance at the local level is not technical or financial. Intuitively it appears that a high level of development is associated with a high level of financial and technical resources. Therefore, the best way to solve this is by the provision of finances and technical expertise. This approach has failed in many cases. Failure of this approach does not mean that technology and finances are not needed. It means that imported technical and financial solutions do not offer appropriate solutions for local problems. Local knowledge, local expertise, and local finances combined with a global support system can help in designing low-cost and affordable solutions by building on the local resource base. This helps communities to cross financial barriers without depending on global loans. A partnership based on the strengths of all the parties helps everyone leverage their resources. It helps communities achieve more without asking for more spending.

As already mentioned, global spending practices are low cost for donors but high cost for local communities. We see similar paradoxes with many other development practices. For example, consider accountability and management. Global projects have to create and integrate accountability mechanisms for ensuring aid effectiveness. Accountability is established through result-based management (RBM). RBM has been developed to counter a bureaucratic input-based approach. However, bureaucratic solution for dealing with the lack of results on ground is to show project money spent as a result achieved, which is another way of showing inputs as outputs.

On the contrary, what appears to be time-consuming and a lack of achievement in a strict timeframe is the unpredictable, zigzag, and slow community mobilization process; it's a process that appears to be out of sync with the RBM but delivers larger, sustainable, and meaningful results over a relatively long period. While intuitively RBM appears to ensure aid effectiveness, it is a counter-intuitive investment in the mobilization *process* that delivers results. RBM is also considered an effective measure for preventing corruption. In the community mobilization process, community engagement makes the whole spending process transparent and an effective antidote to corruption.

Field experience shows that investing in development processes and community endowments ensures cost-effective and sustainable solutions for achieving globally inspired goals. It strengthens community ownership and provides sustainable financial solutions by generating a continued income stream for carrying out local projects. However, this approach does not appeal to the donors because they have limited trust in local partners and consider lack of control over local partners as lack of accountability. This again reflects the trust deficit between global finance and local communities, camouflaged as a technical and financial issue. Justification of external control in the name of technical and financial oversight is not in line with the fact that local initiatives are organic, and accountability is built in the way communities work. Communities perform effectively without outside control because people are not carrying out these projects for others but for themselves. External financial assistance can generate better spending by investing in mechanisms strengthening trust with the local communities.

Poor communities also help pay back the national debts to international financial institutions (IFIs) through regressive and unjust indirect taxes levied on them. Lending more to the governments does not help the poor communities; it adds to their burden. They would prefer to pay a user fee for well-designed local services rather than borrow expensive solutions. Their capacity and willingness to pay a user fee is confirmed if we look at their opportunity cost for not accessing well-designed services from the public sector. In the absence of services provided by the government, they have to access services at a much higher cost from the informal economy. Therefore, communities are willing to pay for cost-effective services. This also points to the need for better spending on a fee-based service provision rather than more spending on unsustainable service delivery. However, the question remains whether the ill-conceived allocation of resources undermining the delivery of global goals should be addressed through confrontation or engagement. Our examples tell us that lack of access to domestic resources is caused to a large extent by the trust deficit, which feeds on the knowledge deficit. This calls for investments in the development of social infrastructure and the local body of knowledge – that is, social capital. This takes us to another dimension of spending, relevant spending. We think that better spending and appropriate spending are two forms of more spending which are little understood and emphasized. Experiments by development practitioners in very diverse settings, from democratic societies to fragile states, affirm the message that investment in social capital is the best form of spending for better results.

People need to work with experts to develop solutions in line with their social reality. It is important to note here that technical and financial assistance cannot succeed without changing the social relationship between the citizens on the one hand and the state and market on the other. This brings to attention another counter-intuitive solution, the priority of social over financial capital. Social capital accumulation not only improves the achievement of development targets, but it also diminishes the need for external resources and provides a mechanism for sustainable delivery of better services by putting in place solutions that are financially affordable to low-income communities. Financial resources do make the difference if invested in creating social capital because the degree of accumulation of social capital determines the effectiveness of financial and technical means for better delivery of services at the local level (De Soto, 2003).

All the case studies in this book have described the most innovative approaches to develop social capital to enable citizens to re-appropriate the space encroached upon by the state or market. The processes designed for building social capital have not only unleashed the development potential of local communities, but they have also helped them form a new partnership with the state, helping them move from the discourse of patronage to discourse of participation; from the discourse of fear to the discourse of hope. Strategies used by different actors are defined by the local context, but there are certain elements common in all the initiatives taken across the Global South. These initiatives build community power through very simple and easy to implement steps, but they have been ignored by dominant donor and policy circles. The purpose of presenting these cases is to share practical insights with the policymakers. This brings to our attention the third most important counter-intuitive lesson: time-consuming, unpredictable community engagement processes deliver better than professionally designed, time- and target-bound result-based frameworks. This happens due to the social capital created by the process-based approach.

Social capital: the missing link for adapting and achieving global goals

Formal sector and informal sector partners are endowed with different types of resources, and these differences provide the foundation for partnership and better delivery with limited resources. These differences also cause confusion and mistrust, as interdependence is very often perceived as a form of dependence. It is important to mention here that communities are better endowed with land[1] and labour, and government or private sector with capital and managerial skills. This difference is seen as a deficiency by conventional development experts, and the solution is sought by handing out resources rather than leveraging the resources through partnership building. Development practitioners across the globe have discovered very innovative ways of using micro-assets for achieving macro results through partnership development. This brings home the point that spending better is spending more. Some of the strategic community-based approaches include *component sharing, incremental financing,* and creating *economies of scale.* The list is very long, but for the sake of demonstrating the strategic impact of local level partnerships with small asset owners, I will briefly explain the concepts mentioned above. This will bring home the point that small is not only beautiful as hinted by Schumacher but strategic as well (Schumacher, 2011).

The concept of component sharing is very simple. According to this concept community-level development consists of two components, internal and external. Internal development has to do with the construction of physical assets and the management of services at the lane and household level; external development has to do with everything beyond the household and lane. Under the component-sharing approach, the government takes the responsibility for external components and the community for the internal component. This approach has been used for the provision of water and sanitation services, solid waste management, and primary education. It reduces the demand on government resources and makes possible achievement of service provision at a larger scale with fewer resources.

The concept of economies of scale has provided foundations for the expansion of the microfinance sector across the globe. While lending to microentrepreneurs incurs losses to the private sector due to its extremely high administrative costs, community savings and borrowing create "economies of scale," as a community organization rather than the individual borrower serves as a "legal person." This makes provision of credit to low-income households feasible as the community internalizes the cost of loan disbursement. Incremental development is another concept which has been built on the power of "less is more." It has been very effectively used in providing shelter to the poor. Here again, the insights gained from the informal sector have helped design a strategy based on the concept of better spending. A fully developed settlement increases the cost of a piece of land, and it becomes out of the reach of the poor. Under the incremental development concept, house-building and development of services are both carried out in phases, and occupants pay as services are developed. This reduces the cost of development as well as the size of the down payment and the instalments to be paid by the homeowner. These approaches succeed due to close interactions between the service providers and the communities and the prevalence of a high level of trust between them (Siddiqui, 1994).

Policy advice: do the right thing on a modest scale, and you will see the snowball effect

Understanding communities and interpreting their responses is a challenging task. Here again, counter-intuitive ways of thinking are extremely helpful. Leading development practitioners have noted a state of inertia in unserved or underserved low-income communities due to a knowledge deficit of government and low social capital of communities. Inertia here means that availability of required resources and unmet needs does not lead to local action for improvement of human development indicators. Limited understanding of the art of working with communities perpetuates this inertia. It is therefore important to benefit from the insights of highly experienced development practitioners in dealing with this problem. These insights are very simple and clear. One, if people don't agree with you it does not mean that they are rejecting you; they are unsure of your ideas. Don't try to convert them through awareness-building; look for early adopters who share your view and are willing to test your ideas. If early adopters succeed, others will follow. Two, when you search for solutions, differentiate between people's dreams and the solutions that are possible within their means. Show them the way to develop solutions within their means, and they will cooperate. Three, always start small to avoid large-scale disasters and loss of trust. Test and develop your model and demonstrate. When you succeed, dissenters will follow you. Four, let communities expand your work through self-selection, and you will see the snowball effect. Ernesto Sirolli echoed the same idea by saying that if you want to help, shut up and listen; and be worthy of invitation before you offer your help (Sirolli, 1999). Mao Zedong used a different expression for conveying this idea: a single spark can create a prairie fire (Mao, 1953). That is what the recipe of better spending is all about.

Note

1 Hernando De Soto in his groundbreaking work *Mystery of Capital* states that value of land used by squatters is worth more than the value of stocks in all the leading stock exchanges in the world. This asset is locked because it does not carry a legal title. It therefore turns into dead capital. The experience of working with people in low-income communities has demonstrated that people are willing to donate land, or to lend the use of a room for starting a home school or a clinic. Or they are open to the use of physical assets for local development. Similarly, they generously offer voluntary labour. They only need technical and social guidance and seed capital in most of the cases.

References

De Soto, H. (2003) *The Mystery of Capital: Why Capitalism Triumphs in the West and Fails Everywhere Else.* New York, Basic Books.

Freire, P. (1970) *Pedagogy of the Oppressed.* New York, Bloomsbury Publishing.

Groucher, S. L. (2004) *Globalization and Belonging: The Politics of Identity in a Changing World.* Oxford, Rowman and Littlefield.

Kahneman, D. (2011) *Thinking, Fast and Slow.* New York, Allen Lane.

Mao, Z. (1953) *A Single Spark Can Start a Prairie Fire.* Beijing, Foreign Languages Press.

Radjou, N., Prabhu, J., & Ahuja, S. (2012) *Jugaad Innovation: Think Frugal, Be Flexible, Generate Breakthrough Growth.* Hoboken, NJ, John Wiley & Sons.

Schumacher, E. F. (2011) *Small Is Beautiful: A Study of Economics as If People Mattered.* New York, Random House.

Siddiqui, T. A. (1994) Shelter for the urban poor, in Siddigui, T. A. (ed.), *Developmental Issues: Innovations & Successes*, Karachi, Fiction House, n.p.

Sirolli, E. (1999) *Ripples from the Zambezi: Passion, Entrepreneurship, and the Rebirth of Local Economies.* Gabriola Island, BC, New Society Publishers.

Stephen, W. (2004) *Think Global, Act Local: The Life and Legacy of Patrick Geddes.* Edinburgh, Luath Press.

Wright, R. (2001) *Nonzero: The Logic of Human Destiny.* New York, Vintage.

Appendices

Appendix A

Area assessment district coverage by province

Badakhshan programme region	1. Arghanjkhawh
	2. Baharak
	3. Maimai
	4. Nusai
	5. Ishkashim
	6. Jurm
	7. Khash
	8. Kufab
	9. Shughnan
	10. Shukai
	11. Wakhan
	12. Yamgan
	13. Zibak
Takhar programme region	1. Baharak
	2. Farkhar
	3. Kalafgan
	4. Warsaj
Baghlan programme region	1. Banu
	2. Dahna-e-Ghori
	3. Deh Salah
	4. Doshi
	5. Khinjan
	6. Markaz-e-Aybak
	7. Khuram Wa Sarbagh
Bamyan programme region	1. Kahmard
	2. Panjab
	3. Shibar
	4. Waras
	5. Shaikh Ali
	6. Surkh Wa Parsa

Appendix B

PAMPI indicators with missing values

Dimension	Indicator	Missing values (including inapplicable and doesn't applied) N = 3,300 Households	Percent missing N = 3,300
Food security	Food security	0	0.0%
Employment	Employment	0	0.0%
Health	Maternal and child health	20	0.6%
	Prenatal care	20	0.6%
Education	Year of schooling	0	0.0%
	Children enrolment	0	0.0%
Living standard	Electricity	5	0.2%
	Sanitation	13	0.4%
	Water	5	0.2%
	Cooking fuel	13	0.4%
	Heating source	12	0.4%
	Asset	2	0.1%
	Floor cover	2	0.1%
	House ownership	0	0.0%

Appendix C

Levels of poverty for different cut-offs

Deprivation cut-off	Level of poverty N (households) = 3,265
10%	98.7%
20%	94.1%
30%	85.8%
40%	**71.6%**
50%	52.6%
60%	33.0%
70%	13.5%
80%	6.2%
90%	1.3%
100%	0.0%

Index

Note: Page numbers in **bold** indicate tables; page numbers in *italics* indicate figures.

Afghanistan: benefits of National Solidarity
  Programme (NSP) to 139–143; Citizen's
  Charter Project 143–146; origins of NSP
  in 137–138; process of NSP in 138–139;
  Programme Area Multidimensional
  Poverty Index (PAMPI) in 150–153;
  state-building in 135–137
aid effectiveness: impediments to 173; and
  local mobilization 106–107, 129, 222–223;
  and spending 44–46, 54–55
aid efficiency 150–153, 161
Andhra Pradesh, India: and continuity of
  leadership 76–77; government in 66–67,
  75; programme replication in 65–66;
  and social entrepreneurship, 64–65
Anjuman Samaji Behbud (ASB): business
  model 98–99; operations 101–104

Bangladesh: benefits of BRAC to 85–88;
  birth of 78–79; changing goals of 88–90;
  and Peru 44–46, 50–54, **51**; poverty in
  82–84; SDG tracking 46–48, 54–57;
  social mobilization in 80
Bangladesh Rural Action Committee
  (BRAC) 10; better spending principles
  85; methods of 81–84; origins of 78–80;
  results of 85–90
Bhalwal, Punjab, Pakistan, 94, 98–102, *102*

Canada: aid to Afghanistan 143–144; First
  Nations as Global South 15, 215; SDG
  data hub 46
Cape Town, South Africa 206–212
Citizens' Charter Afghanistan Program
  (CCAP): origins of 139; purpose of
  143–146

civil society organizations (CSOs) 6–8,
  13–14; heterogeneity of 12; in water and
  sanitation in Pakistan 94
collaboration theory 45–46, 49, 56
communicative rationality 48–50
community-based organizations (CBOs) 96,
  107–109, 112–116
community development: in Bangladesh
  79–81, 90; in India 95–98, 98–99
Community Development Councils
  (CDCs): benefits of 139–143, 145–146;
  origins of 137–138
conscientization (concept) 79–80, 83–85, 88
cost-benefit analysis 30
country ownership 21–24

debt 22–23, 34, 114, 116n2, 222
Democratic Republic of Congo (DRC)
  176–178, *178*, 181
development cooperation 43–44;
  see also South–South Cooperation

Earth Summit 4
equality *see* gender equality
Ethiopia: famine in 184–186; food aid in
  188–192, *193*; food security challenges
  in 186–187; food security policies of
  187–188; sustainability of food sector in
  193–195

First Nations 15, 215
Flying Visit (strategy) 95–97
food security: aid for 188–192, *193*;
  challenges to 186–188; sustaining
  193–195
Freire, Paulo 79–80, 85, 88, 220

gender equality 27, 30, 152; through CDCs
in Afghanistan 141–142, 144–146; lacking
in sub-Saharan Africa 171, 181, 187,
208–209; as MDG 3 141; principles of
BRAC 83, 85–86, 89; as SDG 5 86,
167; *see also* women
governance: failures of 120, 176; global
experiments with 136–138; and water
management 97–101, **175**, 210;
see also new public governance (NPG)

human development: in Bangladesh 83, 85;
relation to financial resources 4–6, 21;
relation to government 11–12
Human Development Index (HDI) 50–51,
51, 114
human development indicators (HDIs)
4–5, 50
Human Development Report 149–150

India: government of 66–67, 77n2;
RAS/96/600 project in 76–77; SAPAP in
62, 74–75; social mobilization in 64–65,
65; *see also* Andhra Pradesh
indigenous peoples 15, 215
inequality *see* gender equality
international financial institutions (IFIs):
and debt 113–115, 222; and The World
Bank 19–21, 28, 33–35

Karachi, Sindh, Pakistan: and Orangi Pilot
Project 110–116, 116n3, 116n4, 116n6;
urban context of 107–109

Local Agenda 21 4

Memorandum of Understanding (MOU):
Anjuman Samaji Behbud (ASB) 98,
102–103; Bangladesh-Peru 44–46,
51–56
microcredit: and Bangladesh Rural Action
Committee 81–86; and Orangi Pilot
Project 116n6, 123–125
microfinance: approach to 124–128; in
developing markets 129–132; growth of
119–123, **123**; and women 86, 89
Millennium Development Goals (MDGs),
4, **154**; in Afghanistan 138, 140–141,
152–153; in Bangladesh 85–86, 88;
and Bangladesh-Peru memorandum
46–47; in sub-Saharan Africa 166–167,
167; *see also* Sustainable Development
Goals (SDGs)

National Solidarity Programme (NSP):
benefits of 139–143; origins of 11,
135–137; process of 138–139; purpose
of 28–29, 137–138; *see also* Citizens'
Charter Afghanistan Program (CCAP)
new public governance (NPG) 45–46,
48–49, 57n3
non-governmental organizations (NGOs)
4, 21, 72; in Bangladesh 78–80, 88;
funding of 41n7, 66, 126; in India
64–65, 68–69; in Pakistan 116n3,
122, 125; *see also* community–based
organizations (CBOs)

Obras Sanitarias del Estado (OSE): history of
201–202; as model public water company
204–206; role of workers in 202–204
Official Development Assistance (ODA): in
Democratic Republic of Congo 176–177;
and outcomes in sub-Saharan Africa (SSA)
179–181; in Pakistan 129; in SSA 166,
169–176, *171*, *174*, **175**, *175*
Orangi Pilot Project (OPP) 94, 101; origins
of 107–108, 116n5; purpose of 110–112,
125; Research and Training Institute
(OPP-RTI) 112–114, 116n3, 116n6
Orangi Pilot Project Research and Training
Institute (OPP-RTI) 112–114, 116n3,
116n6

Pakistan 31–32; aid financing in 26,
105–107, 113–115, 120–123; and
assessment instruments 31–32, 116n3;
birth of Bangladesh 79; market
constraints in 130–132; microfinance
approach in 10, 124–130; poverty
in 29, 116n1, 116n2; Rural Support
Programmes in 62–64, 75; urban
planning in 115–116; water sector in 9,
93–94; *see also* Karachi; Punjab
Pakistan Poverty Alleviation Fund (PPAF):
origins of 121–122; role of 125–126
participatory initiatives 9–10, 153; in
Afghanistan 136–137, 144; in Bangladesh
80; in India 73–74, 95; in Pakistan 112–113
policy transfer 44–46; Bangladesh-Peru
50–52, **51**, 56–57; and new public
governance (NPG) 48–50; South-South
52–53, 55–56
poverty alleviation: civil society's role in
10; as global goal 4, 28–29; jargon of 12;
as local goal 5–6; in South Asia 61–66,
65, 75–77; *see also* PPAF; PAMPI

Programme Area Multidimensional
Poverty Index (PAMPI): measuring
152–155, **154**, **155**, **229**; purpose of
150–152; results of **156**, 156–160, *157*,
158, **159**, *160*; use of 161
Punjab Health Engineering Department
(PHED) 95–99, 101
Punjab Municipal Development Fund
Company (PMDFC) 99–100
Punjab, Pakistan: aid financing in 99–102,
124; and assessment instruments 31;
participation strategy in 95–98, 104n2;
politics in 94–95, 106

right-based development 6, 12, 94

sanitation 13, **155**, 157, 229; in Afghanistan
146, 153, *160*; in Democratic Republic
of Congo 176–178, *178*; disparities in
sub-Saharan Africa (SSA) 165–167;
improvements across SSA *168*, *169*,
179–181; and MDGs **154**, 166; in
Pakistan 94, 101–103; and Pakistan's
Orangi Pilot Project 108–112, 116n5,
116n6; and Pakistan's Urban Resource
Centre 112–114; in SDG 6 86, 140–141,
144; services in SSA 170–174, *171*,
173–176, **175**; in Uruguay 200–206,
211; *see also* water management
seed capital 66–68, 70–74, 225n1
SFA (SMEC Foundation Afghanistan)
150–153; PAMPI use in 161;
programme regions 159
SMEC Foundation Afghanistan (SFA)
150–153; PAMPI use in 161;
programme regions 159
social capital 8–12, 136–137, 222–224
social development 6–11, 64, 81, 130–132;
see also human development
social entrepreneurship 64–66
social movements: in Cape Town 203,
209–211; in Uruguay 201–203
Society for Elimination of Rural Poverty
(SERP) 61, 65, 75
South Africa *see* Cape Town; sub-Saharan
Africa
South Asia Poverty Alleviation Programme
(SAPAP): birth of 62–63; concerns
about 73–75; strategy of 64–66

South-South Cooperation (SSC):
mainstreaming of 55–57; methodology
45–46, 57n1; objectives 43–45; policy
transfer 48–54; timing of 52–53
spending: better models for 6–7, 98–104,
124–130, 142–143; better principles for
85, 88; problem of 11–13; role of local
actors in 108, 219–225
sub-Saharan Africa (SSA): Democratic
Republic of Congo 176–178; and
development goals 166–169, 171–176;
investments in water and sanitation
169–170, *171*, *173*, *174*, **175**, *175*;
water access across *167*, *168*, *169*,
179–181
Sustainable Development Goals (SDGs) 4,
176; in Afghanistan 138, 140–142, 144;
in Bangladesh 85–88; and Bangladesh-
Peru memorandum 44–46, 48–53; in
sub-Saharan Africa 166–167; tracking
46–48, 54–55; *see also* Millennium
Development Goals (MDGs)

trade unions 202–206

UN Office for South-South Cooperation
(UNOSSC) *see* South-South
Cooperation (SSC)
Urban Resource Centre (URC) 12–13;
origins of 107–108; purpose of 112–115
Uruguay 201–206, 211–212

water management: and access across sub-
Saharan Africa 166–167, *169*, 180–181;
and access in Democratic Republic
of Congo 171–172, *178*; and access
in Pakistan 101–103; and governance
97–101, **175**, 210; in-home devices for
207–211; and politics 94–95; OSE as
model public water company 204–206;
and wastewater 109–113; *see also*
sanitation
women: and CDCs in Afghanistan 140–145;
and poverty in Pakistan,106, 121–124,
123; and purpose of BRAC 81–85; and
SDGs in Bangladesh 85–89; and self-
help groups 64–66, **65**, *68*, 69; and water
collection in sub-Saharan Africa 167–171;
see also gender equality

Printed in the United States
by Baker & Taylor Publisher Services